# THE AMATEUR
# ASTRONOMER

# THE AMATEUR ASTRONOMER

BY ANTONÍN RÜKL

CONSULTANT EDITOR
JOHN GRIBBIN

Translated by Olga Kuthanová
Graphic design: Pavel Büchler

English version first published 1979 by
Octopus Books Limited
59 Grosvenor Street, London W 1

© 1979 Artia, Prague

ISBN 0 7064 1034 3

Printed in Czechoslovakia
3/99/36/51-01

# CONTENTS

# FOREWORD

*With interest in astronomy at the amateur level probably running stronger than ever before in Britain, this translation of Rükl's book is particularly timely, and I am pleased to be associated with the English text. This is very much a handbook to aid observations of the night sky — even if they are only casual observations made with the unaided human eye — and because of this perhaps a word of guidance on how best to use the book is appropriate. Quite rightly, the author moves from a general background discussion setting the scene of man's place in the Universe to the detailed pictorial section of immediate practical use as an aid to observations. The details of such essential but less exciting aspects as choosing a telescope are left to the final part of the book, although in practice, of course, you will need some observing instrument in order to make best use of the pictorial section. For the complete newcomer to astronomy, then, don't try to plunge straight in and find the details described in some of the plates. Read the book once for flavour, then with the advice from the last section decide on an instrument to use and a programme of study. At that point it will be time enough to return to the details of the pictorial section and begin a lifetime involvement with the wonders of the night sky.*

*John Gribbin*

# PREFACE

All around us, as far as the eye can see, is the Universe. The planet Earth is part of this Universe and so are we. It is our large, infinite world. At first sight it seems strange, mysterious, inaccessible, incomprehensible. But we can get to know it and by doing so enrich our lives by seeing both our surroundings and ourselves with new eyes.

This book is addressed to all friends of nature. It is a small astronomical atlas and illustrated guide to interesting objects and phenomena that can be observed by means available to the amateur. 'Our' Universe — man's Universe — is the local region of space that we can observe by such means, dominated by the Sun and its family of planets and other objects — the Solar System. In the infinity of space even this region is small and insignificant. But to man, living on planet Earth, it is home.

Beginners who have no previous knowledge of astronomy will find in the introductory text information about certain basic terms which will prove useful for easier orientation in the pictorial section. The final chapter contains instruction for simple observations. In between, the pictorial section begins with Earth, Moon and Sun, and moves on to other objects in the order of their distance from the Sun. The selection of objects and manner of depiction will enable the reader to find suitable subjects for observation with the naked eye, binocular and telescope. It also shows in brief the structure of the Universe. Sky maps are provided for general orientation, and maps of the Moon, Mercury and Mars (each with index of named features) may be used by the more advanced amateur before he moves on to the more detailed maps or atlases of these celestial bodies used by the more dedicated astronomical hobbyist.

Most of the plates show the celestial bodies as we see them from Earth. Their coloration is mostly exaggerated for effect, and in several places a schematic view as it would appear to an extraterrestrial observer is also included in order to better explain a certain phenomenon. A book of this kind can only provide a superficial picture of the surrounding world, but it will have served its purpose if it awakens the reader's interest in learning more about the Universe; for the guidance of such readers, some suggestions for further reading are provided on p. 17.

The author extends his sincere thanks first and foremost to Hermann Mucke, Director of the Vienna Planetarium, for his invaluable technical assistance in the initial stages of work on the illustrations and in particular for supplying the initial data for constructing the graphic tables showing the positions of the planets. Thanks to Jean Meeus, renowned Belgian astronomer, who kindly permitted the use of his computations, the author was able to augment the given tables with reliable data on the phases of the Moon and aspects of the planets.

Compiling charts of the Moon, Mercury and Mars in line with current knowledge of these bodies is impossible without modern catalogues, mapping material and photographs made by space probes. This material was willingly made available to the author by the American astronomers Dr. Merton E. Davies of the Rand Corporation, California, Dr. Leonard Jaffe of the Jet Propulsion Laboratory, California, and Leif Andersson and Ewen A. Whitaker of the Lunar and Planetary Laboratory, Tuscon, Arizona, to all of whom the author extends his sincere thanks.

He is also indebted to Prof. Vladimír Guth, Dr. Sc., of the Astronomical Institute of the Czechoslovak Academy of Science, and Ing. Pavel Příhoda of the Prague Planetarium, for reading the manuscript and making valuable reminders.

Last but not least the author extends his thanks to his wife, Sonia, for all she did to facilitate his work on this book and complete it on time.

# THE EARTH IN SPACE

From childhood we learn about Earth as a planet moving through space, but let us be honest: can we really perceive this space around us? Are we not in the habit of seeing ourselves 'down' on the Earth looking 'up' at the sky — that giant vault which is azure-blue in the daytime and studded with stars at night?

Clouds, the rainbow, the northern lights, the Sun, Moon and stars all appear 'in the sky', in an immensely high dome reaching from horizon to horizon. How we view the sky depends on us. Will we be content with the idea of a vault above the landscape or shall we try for a spatial perspective?

In the first instance we can view the sky as a large natural planetarium with a non-stop programme. In nature, just as in a regular planetarium, we easily succumb to the illusion that the dome-sky revolves during the day from east to west carrying the Sun, Moon and stars with it. The most striking difference between the two is size. The hemispherical domes in the largest planetariums measure 20 to 25 metres in diameter, whereas the diameter of the natural 'celestial dome' is infinite. In astronomical terminology it is called the *celestial sphere*. It is on this imaginary sphere of infinite diameter that we project all celestial objects without regard to their distance.

Orientation in the celestial sphere is based on the four cardinal points, which can always be easily determined by one of several common methods (by means of the Sun and a watch, the compass or the North Star). Knowing the *constellations* is a great help in finding one's way about the heavens. The figures outlined by the stars, and representing various mythological beings, animals and objects, are products of man's imagination, and those in the Northern sky were given names thousands of years ago. The randomly scattered stars in the celestial sphere afford numerous possibilities for connecting them to form the outlines of various images or figures

as the evidence of imaginative maps and globes of olden times testifies. Present-day constellations are usually limited to groups of bright stars delineating simple geometric figures. The imaginary lines connecting the stars and making it easier to remember the constellations are called *alignments;* they may be found in the constellation maps on pp. 140—157. More detailed maps usually show also the complex, internationally valid boundaries of the constellations, which number 88 all told.

Every amateur astronomer should know at least the most important constellations such as the Great Bear, Cassiopeia, Orion and the Swan. Knowing the 12 zodiacal constellations (Pisces, Aries, and so on) may also prove useful; they provide a background against which we can then easily identify the planets and observe their movements. The heavens are much more interesting if one can find one's way about there.

For precise determination of positions in the celestial sphere astronomy makes use of several types of spherical coordinates. Figure 1 at the left presents a diagram showing the relation between the system of coordinates on the Earth and in the celestial sphere. Let us start with the Earth's axis of rotation passing through the North Pole $N$ and South Pole $S$; the points at which this axis extended intersects the celestial sphere are the *north celestial pole $P_N$* and *south celestial pole $P_S$,* respectively. Through these poles passes the axis of the celestial sphere round which the latter seemingly rotates from east to west in the opposite direction to that of the Earth's real rotation.

Parallel to and above the terrestrial equator a' extends the *celestial equator* a. Similarly we can endow the celestial sphere with meridians and parallels corresponding to the system of geographical coordinates, which brings us to the system of coordinates that are of fundamental importance in astronomy — the *equatorial coordinates.* It is nec-

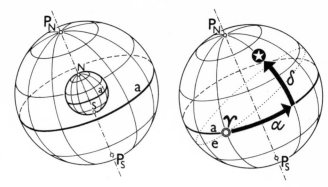

Fig. 1. Equatorial coordinates

essary to point out that the Earth's dimensions are negligible compared with those of the celestial sphere, so that in Figure 1 the Earth can be reduced to a mere pin-point, identical with that marking the location of the observer. Equatorial coordinates are universally valid, and star maps based on this system may be used anywhere on the Earth.

We have already pointed out that equatorial coordinates in the celestial sphere are analogous to the geographical coordinates. Accordingly, we may say that the equivalent of 'Greenwich' (i.e. the point through which the prime meridian passes) in the celestial sphere is the *vernal equinox* ♈, the point where the celestial equator a intersects the ecliptic e (Figure 1 at right). The ecliptic is the apparent annual path of the Sun on the celestial sphere which passes through the 12 zodiacal constellations. In the sky *right ascension* α is the equivalent of geographical longitude on the Earth; it is measured eastward from the vernal equinox and expressed in degrees (from 0 to 360) or, more commonly, in hours (from 0 to 24).

*Declination* δ is analogous to geographical latitude; it is measured in degrees positive (+) for angles north of the equator and degrees negative (−) for angles south of the equator, each from (0° to 90°). So specifying right ascension and declination specifies a unique point on the celestial sphere — a map reference in the sky.

At a given instant we can see only the part of the celestial sphere that is above the horizon of the place of observation. In practice it is often useful to express the positions of celestial objects in terms of so-called *horizontal coordinates*. We can easily picture them if we know how to locate the four cardinal points and learn to estimate angular di-

mensions — horizontal and vertical angles. The horizontal angle measured from the south point on the true horizon westwards is called the *astronomical azimuth*. The second horizontal coordinate is *altitude* (above the horizon); this is measured from 0° on the true horizon to 90° at the zenith. The figures on page 27 showing differences in the duration of day and night at various geographic latitudes are depicted in horizontal coordinates. Orientation usually starts with the *meridian* — i.e. the great circle of the celestial sphere passing through the celestial poles and the zenith.

To get to know the stellar sphere better, try for a spatial picture. For a start it will suffice to divide astronomical objects into those that are closer and those that are more distant. Leaving aside meteors and artificial earth satellites close to our planet, celestial objects may be listed according to their distance from the Earth in the following order: Moon; planets, Sun and comets; individual stars; Milky Way; galaxies. Such a division is easy in terms of relative distances; difficulties arise, however, when we try to apply a scale of measurements. Kilometres will suffice for expressing distance in the Solar System, but even there it is often more advantageous to use so-called astronomical units. An astronomical unit (A.U.) is a unit of length equal to the mean distance of the Earth from the Sun, about 150 million kilometres (precisely 149,597,870 km). For example, if we say that Jupiter is 5.2 A.U. distant from the Sun, it means that this planet is 5.2 times farther from the Sun than the Earth is.

Distances in space are so far beyond our experience that usually we are unable to imagine them in absolute terms. This, however, need not bother us unduly, because we can

9

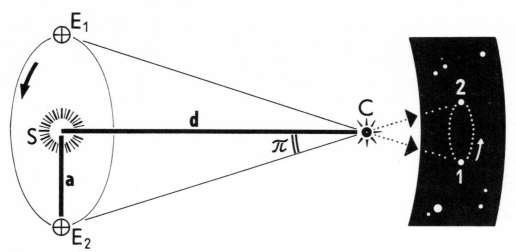

*Fig. 2. Annual parallax*

compare the relative distances of celestial objects and that is very valuable for obtaining a correct idea of their distribution in space.

In popular and science fiction literature we often come across the term *light year,* a unit of distance equal to the distance that light travels in a vacuum in one year. The speed of light is 300,000 kilometres per second, so we can easily deduce that one light year is equal to a distance of approximately 9.5 billion*) kilometres ($9.5 \times 10^{12}$ km). For example, the star Alpha Centauri is 4.25 light years distant, star cluster M 13 in Hercules is about 25,000 light years distant, and from galaxy M 31 in Andromeda it takes light more than 2,000,000 years to reach the Earth. The Sun, on the other hand, is only 8.5 light minutes distant and the Moon a mere 1.3 light seconds.

In specialist literature interstellar distances are usually expressed in units of measure called parsecs. A *parsec* (pc) is the distance from which one astronomical unit appears in one second of arc ($1''$). Parsec is more clearly explained by Figure 2, which shows the trigonometrical method of determining distance $d$ by means of annual parallax $\pi$. If we observe the near star C when the Earth is in position $E_1$, the star is located at point 1 against the distant stellar background. In half a year the Earth will move round the Sun S to position $E_2$ and the star will be at point 2. If we can measure the extent of the parallactic dis-

placement 12 (the angular distance between points 1 and 2), then we can derive the annual parallax $\pi$, which is the angle subtended at the star by the radius of the Earth's orbit — the equivalent of astronomical unit $a$. Thus:

$$d = 1 : \pi$$

with $d$ expressed in parsecs if $\pi$ is expressed in seconds of arc.

A star 1 parsec distant would have a parallax of $1''$; however, there is none as close as that in the Sun's neighbourhood. The Sun's closest neighbour is the triple star Alpha Centauri. This has a parallax of $0.75''$, which corresponds to a distance of 1.3 pc (i. e. 270,000 A. U.). Most stars visible to the naked eye are within a range of several tens to several hundreds of parsecs. With a telescope we can observe objects in the Milky Way, some of which are thousands and even tens of thousands of parsecs distant. Within our Galaxy distances are expressed in terms of kiloparsecs (1 kpc = 1000 pc); in the world of outer galaxies they are usually expressed in megaparsecs (1 Mpc = $10^6$ pc = 1,000,000 pc).

Copernicus (1530) estimated the distance of the stars to be about 4,000 times the distance from the Earth to the Sun. Kepler (1620) raised this estimate to 10,000 and Newton (1700) to 20,000 astronomical units. The first measurements of stellar parallaxes — a technique depending on the apparent movement of stars as the Earth moves round the Sun — in the 1830s showed, however, that the nearest stars are several hundred

---

*) Throughout this book one billion = one million million.

thousand to a million times farther from the Earth than the Sun is, thus opening to man the breathtaking concept of infinite space where the light of innumerable stars and stellar systems shines like beacons.

Measuring the distances of stars opened the way to learning about their characteristics. The first and most obvious is their brightness. According to tradition, established by Hipparchus in the second century B.C., stars were originally divided into five groups of *stellar magnitudes.* The brightest stars, those that are the first to appear in the sky when darkness falls, were classed as stars of the first magnitude and the faintest stars still visible with the unaided eye as stars of the fifth magnitude. It was believed that the bright stars are larger than the faint ones; in reality the matter is far more complex, but still 'stellar magnitude' continues to be the term used to express the intensity of stellar brightness. The present-day scale differs from Hipparchus' in both range and definition. The faintest stars visible without a telescope are of the 5th to 6th magnitude. With an ordinary telescope we can observe stars of the 10th to 13th magnitude and can photograph even fainter objects. With the largest astronomical telescopes and best photographic aids it is possible to photograph stars of up to approximately the 24th magnitude. According to the scale, one degree of magnitude corresponds to a difference in brightness of 2.512 times. For example, a star of the 3rd magnitude is 2.5 times fainter than a star of the 2nd magnitude. The range of five stellar magnitudes corresponds to a brightness ratio of 1:100 ($2.512^5 = 100$). At the other end of the scale are the bright stars. For example, the brightest star of the summer sky — Vega in the constellation Lyra — is of 0 magnitude and the stellar magnitude of even brighter objects is expressed by negative numbers — Sirius $-1.6^m$, Venus as much as $-4.6^m$, the full Moon $-13.6^m$ and the Sun $-26.8^m$.

These magnitudes are termed apparent magnitudes and are designated by the small letter m next to the number. On star maps stars of different magnitudes are usually represented by discs of differing diameters. In this case the term 'apparent' is most impor-

tant, for a certain star may appear to be very bright simply because it is near to us, whereas another may be barely visible because it is very distant.

A better criterion for comparing the brightness of stars is their so-called absolute magnitude M, determined by calculations based on a star's apparent magnitude and distance from the Earth and telling us how bright the star would be if it were located at a distance of 10 parsecs from the Earth. If we were to move all the stars in the heavens to the same distance, the sky would be changed beyond recognition. It would be populated with stars brighter than Venus, while the Sun, at a distance of 10 pc, would appear as a small star of the 5th magnitude. Determining the true substance of apparent phenomena always was and continues to be of prime importance in astronomy. Deciphering astronomical observations may sometimes be likened to a detective investigation with a surprising outcome.

Astronomy is unique among the natural sciences in that it studies and learns about distant objects indirectly, mainly through their *radiation.* Of the entire spectrum of electromagnetic radiation, *light* perceived by the human eye is the most important to man. Light rays were the principal and sole bearers of information in astronomy until the 1940s.

*Classical astronomy* was concerned chiefly with determining the direction from which light rays come — in other words, determining the position of celestial bodies for the purpose of studying their movements, measuring time, navigation, and the like.

With a prism or diffraction grating light can be dispersed into a spectrum, into separate components (colours, wave lengths). The clear, rainbow colours in the spectrum are not nearly as interesting to the scientist as the finer structure of the spectrum — the numerous spectral lines that reveal the presence of particular elements and provide much further information. Atoms of a chemical element usually emit energy in whole series of precisely limited wavelengths that occupy specific position in the spectrum — these are the spectral lines. One might compare the spectrum to the dial of a radio set where one tunes in on the various transmitting stations according to given wavelengths or frequencies.

The fine lines in the stellar spectrum tell us not only about the chemical composition of the stars (more or less the same for most stars), but also, what is most important, about the physical conditions of their environment. Spectra are veritable mines of information for *astrophysics* — the modern branch of astronomy concerned with the study of the physical properties and phenomena of the stars, planets and all other heavenly bodies. A star's spectrum and the intensity of its light make it possible to determine the temperature of the star, the velocity of its rotation, the velocity of its approach or recession, the presence of a magnetic field, and indirectly also its dimensions, distance from the Sun, and age. The astronomer does not obtain this information by direct observation with a telescope but by the subsequent measurement and evaluation of photographs of the spectra.

Astrophysical methods of investigating the Universe have been greatly extended with the development of *radio astronomy,* which studies celestial bodies and objects by means of radio waves. This branch is responsible for many important discoveries which are an inseparable part of modern astronomy.

Our knowledge of the Universe would be greatly limited if we were confined to making observations from the surface of the Earth through the dense filter of its atmosphere (see page 24). The Earth's atmosphere is a serious obstacle to astronomical observations. Not only does it distort, depreciate and limit observation in the region of visible radiation, but it also makes it totally impossible to observe the Universe in most other regions of electromagnetic radiation. Transporting instruments above the ground layers of the atmosphere by means of balloons and rockets and above all the development of artificial satellites in the 1960s made it possible to overcome even this eternal obstacle. Now astronomers are coming to grips with entirely new views of the Universe in the region of infrared and ultraviolet radiation, X-ray radiation, gamma radiation, and further developments of observations.

Direct, visual observations are of limited value and astronomers gaze through telescopes only on rare occasions. Pictures obtained through the telescope are generally recorded on a photographic emulsion, so that modern telescopes may be best described as large photographic cameras. For analysis of light, telescopes are fitted with extremely complex and expensive optical and electronic equipment the cost of which is often greater than that of the telescope itself. The equipment of laboratories and workplaces for processing observations is equally important. And then there are the special instruments for space astronomy installed on man-made satellites and interplanetary probes. It would be hard to name a branch of science with greater variety of equipment than present-day astronomy.

These expenditures in terms of equipment and effort have yielded a corresponding influx of new information about the Universe, new discoveries and changes in opinion even in those areas that until recently were believed to be a clear-cut and closed chapter. Our view of the starry sky is thus continually expanded and enriched. From his primitive conception of the vault of heaven man has been brought by astronomy to an understanding and comprehension of the spatial arrangement of the heavenly bodies among which our planet also has its modest place. Richer for the knowledge, we can observe with far greater understanding the phenomena occurring on the stage of the cosmic theatre moving round our observation spaceship — planet Earth.

# STRUCTURE OF THE UNIVERSE

From the viewpoint of the terrestrial observer the arrangement of celestial bodies into units or systems is by no means clearly evident. No wonder, then, that it was not until well into the twentieth century, after man had learned what the galaxies are, that he was able to understand at least in rough outline the structure of the Universe.

If we were to put together a 'cosmic address', for Earth it might be as follows:

Planet Earth
3rd orbit in the Solar System
Star called the Sun
Galaxy 'Milky Way'
Local Group of Galaxies

To this we might add a 'zip-code' composed of the coordinates of the Sun in the Galaxy.

The following is a brief annotation to the address. The Earth is part of the Solar System; it is the third planet in order of its distance from the central body — the Sun. The Sun is an ordinary star, not something exceptional in the Universe. Together with the planets and other bodies of the Solar System, the Sun is a member of a huge stellar island known as a Galaxy. Our Galaxy is called the Milky Way. Galaxies are fundamental building-blocks of the Universe. They form groups and clusters of galaxies comprising tens and hundreds of members. Our Galaxy is part of the so-called Local Group of galaxies, which has some 25 members.

Let us now take a closer look at the above hierarchy and take note of the general features and evolutionary aspects; some details may be found in the text accompanying the plates.

## The Solar System

The Solar System is our home in the Universe and we know it in some detail. Besides the Sun and nine known planets,*) it includes countless other inhabitants, whose *total* mass is very much smaller than the mass of the central star. Seven planets are circled by various natural satellites or moons (36 identified as of 1978). Besides the nine known planets, the Sun is surrounded by thousands of minor planets, or asteroids, moving mostly between the orbits of Mars and Jupiter. The orbits of the planets and most minor planets are located in planes not too inclined to the plane of the Earth's orbit, and that is why we observe the given objects in the sky near the ecliptic, moving 'through' the 12 zodiacal constellations.

The orbits of another group of objects, the comets, are quite different. Some extend from beyond the orbit of the planet Pluto in closer to the Sun than the Earth ever gets. Meteor streams formed by the disintegration of comets, collections of cosmic junk orbiting the Sun together, have similar orbits. The movements of all members of the Solar System are governed by the law of gravity, which applies throughout the Universe.

The star called the Sun, on the one hand, and the planets, on the other, represent two essentially different types of cosmic bodies, differing primarily in mass. Stars are gaseous bodies which are self-luminous, as a result of nuclear processes in the stellar interior. Stars may differ markedly in size; some are much smaller than the Sun, but there are also other, giant stars, with a diameter greater than the orbit of the planet Mars. Their masses, however, are always within the fairly small range of $10^{32}$ g to $10^{35}$ g. A body with a mass of less than some $10^{32}$ g does not have sufficient heat in its interior to initiate a nuclear reaction.**)

Planets are bodies of smaller mass than stars and are not self-luminous. They are cold on the surface and shine only by reflected light.

The planets in our Solar System are di-

*) The object known as 'Chiron', between the planets Saturn and Uranus, is not a planet but a much smaller object, an asteroid.

**) $10^{32}$ is shorthand for 1 followed by 32 Zeros, or one hundred million billion billion.

13

vided according to their distance from the Sun into two groups — the *inferior planets* (Mercury and Venus) and the *superior planets* (from Mars outward).

They are furthermore differentiated according to their physical and chemical properties into *terrestrial planets* and so-called *major planets*. The first include Mercury, Venus, Earth, Mars, and in all probability also Pluto. They resemble one another not only by having smaller dimensions than the major planets but also in chemical composition, mass and solid surface with similar morphological features. Common surface formations on terrestrial planets (and on their moons) are the ring-like walled enclosures called *craters*. Craters testify to strong bombardment by meteoritic bodies in the early phase of the evolution of the planets and their satellites. Besides impact craters (caused by the impact of meteoritic bodies) there are also craters and many other formations of volcanic or tectonic origin. The external appearance of some planets (Earth, Venus, Mars) has also been influenced in great degree by the effects of atmosphere and/or water.

The present chemical composition, mass and other characteristics of planetary atmospheres are the result of lengthy evolution. Mercury has a very tenuous helium atmosphere. The gaseous envelopes on Venus and Mars consist predominantly of carbon dioxide ($CO_2$). On Venus, however, the atmospheric pressure is 90 times greater than on the surface of the Earth, and on Mars the atmosphere on the surface is about as tenuous as on the Earth at an altitude of 30 kilometres.

As regards the chemical composition of terrestrial planets, the heavier chemical elements — oxygen, silicon and iron — predominate. On the other hand, the chemical composition of the major planets, chiefly Jupiter and Saturn, is more like that of the stars, consisting predominantly of hydrogen (75%) and secondly of helium. The major planets are cold on the surface; they are composed of frozen and liquefied gases and are enveloped by an extensive atmosphere consisting of hydrogen, helium, methane and other admixtures. Rapid rotation, pronounced flattening at the poles, dark and light bands of clouds — these are further striking features of the major planets. It has been discovered that Jupiter and Saturn have their own source of internal heat, but this is apparently the result of the planets' gravitational contraction, and not a sign of nuclear reactions going on inside them. Jupiter has a strong magnetic field, Mercury has a weaker magnetic field than the Earth, and Venus and Mars practically no magnetic field whatsoever.

The basic differences between the terrestrial and major planets are related to the formation and evolution of the Solar System. It is now generally accepted that the Sun and planets were formed simultaneously from a cloud of interstellar gas and dust that acquired the form of a rotating disc which then broke up into a great number of balls. The central part of the disc with the greatest concentration of mass became the Sun and the remainder gave rise to the other celestial bodies. This happened some 4.6 thousand million ($4.6 \times 10^9$) years ago.

An idea of the different conditions on the planets may also be obtained from the following data. On Mercury the daytime temperature rises to about 400°C and to an observer on that planet the Sun would appear as a dazzling disc some 2.4 times larger than to an observer on the Earth. At the other extreme, on Pluto the daytime temperature is approximately −200°C and to an observer there the Sun disc would be almost impossible to distinguish without a telescope; it would look like a very bright star and it would be possible to observe the other stars even during the daytime.

## The Stellar Universe

Let us enter the world of fantasy for a while and imagine that we are flying straight from the Sun at the speed of light, 300,000 kilometres per second. Within 8.5 minutes we would fly past the Earth and in 5 ½ hours we would cross the orbit of the planet Pluto. Continuing on our way, we would note a rapid decrease in the brightness of the Sun, which would glow only like a fading star. It would take a full 4.3 years (terrestrial time) to traverse the distance separating us from the nearest star, called Proxima Centauri; it is the faintest component of the triple star Alpha Centauri in the southern sky. The size of the Solar System is negligible compared with interstellar distances, and it is not surprising that we cannot see stars as they really look, as separate discs, even with the largest telescopes.

Astrophysics, however, has at its disposal methods that make it possible to obtain information about the stellar Universe despite the immense distances separating us from the given objects. For example, the spectrum of a star enables us to deduce its surface temperature, as indicated on p. 116. On the basis of its spectrum and luminosity we can find the position of a star in the so-called Hertzsprung-Russell (H-R) diagram (see p. 118). This diagram is of fundamental importance to the study of the characteristics and evolution of stars.

Stars were formed and continue to be formed by the compression of interstellar gas and dust scattered throughout space. After the pressure and temperature in the interior of a developing star have built up sufficiently and it begins to produce radiant energy, the star enters the H-R diagram 'at the right' (i. e. into the region of low temperatures) as a red star. The newborn star moves fairly rapidly to the Main Sequence (p. 118), where it then passes most of its life, releasing energy by burning hydrogen into helium. After the hydrogen has been exhausted as a fuel, the star switches to a different nuclear reaction, abandons the Main Sequence and becomes a stellar giant. This point marks the end of the best years of its life and the beginning of an unsettled old age ending in dramatic changes.

The entire course of evolution is determined first and foremost by the amount of mass the evolving star captures; the greater the initial mass, the more turbulent and brief the star's existence. For example, a star such as our Sun lives a quiescent life on the Main Sequence for about 10 thousand million years. Conversely, stars that had a far greater mass on entering the Main Sequence and glow like white giants some 10,000 times brighter than the Sun deplete the hydrogen in their cores within tens of millions of years, 'burning' it through nuclear reactions which produce helium and other elements. When the hydrogen is exhausted, the star passes to more complex nuclear reactions and burns increasingly heavier elements. The luminosity of a star then changes and varies (see cepheid variables — p. 122). In the final stage, when all the sources of nuclear energy are exhausted, gravitational contraction takes over as the determining factor in the star's further evolution and this inevitably leads to the star's collapse, compression to a very small size and corresponding changes in the physical structure.

In the case of stars like the Sun this final stage is a 'white dwarf' — a star that is a hundred to a thousand times smaller than the Sun but contains as much matter with a density of $10^5$ to $10^8$ g/cm$^3$ (i. e. as much as 100 tons per cubic centimetre!). Astronomers have known about the existence of white dwarfs since 1862, when the American astronomer Alvan Clark discovered a faint companion to the star Sirius. The star Procyon in the constellation of Canis Minor (the Lesser Dog) also has a white dwarf as a companion. Even more extreme forms of compressed matter produce neutron stars (these have a diameter of only 10 to 100 kilometres and density of up to 100 million tons/cm$^3$) and so-called 'black holes', peculiar remnants left by the gravitational collapse of extremely dense stars.*) The entire evolution of a star can nowadays be followed theoretically by mathematical models with the aid of computers. The accuracy of the theory is attested by observation.

Not all stars are solitary, isolated occupants of space. Many evolved as double, triple or multiple stars. *Double stars* or binaries (p. 120) provide a unique source of information about stellar mass, for astronomers can 'measure' the stars quite reliably on the basis of the gravitational effects of the system's two components. Furthermore, it was discovered that in the case of Main Sequence stars (and these are in the majority) there is

---

*) For more about these bizarre objects, see some of the books given as further reading (p. 17).

an important correlation between mass and luminosity, and this has made it possible to estimate the mass of single stars as well. Stars also occur in large associations known as *star clusters.* Open and globular star clusters (p. 124) are very rewarding objects for observations by the amateur.

Let us return to our imaginary flight from the Sun at the speed of light, this time in the direction of the constellation of Andromeda. First of all we would journey several thousand years among stars five to ten light years apart. Then they would become increasingly fewer in number and we would find ourselves leaving the giant stellar system of our Milky Way galaxy. After several tens of thousands of years of flight (still measured in terms of terrestrial time) we would travel beyond the boundaries of our Galaxy and be able to view it in its entirety (see p. 110). After that we would journey hundreds of thousands of years through practically empty space. We would have passed the stars in the Andromeda constellation in the first centuries of our fantastic flight and the concept of Andromeda as a constellation would have lost its meaning even before that; but a nebulous cloud, the well-known galaxy in Andromeda, would still be ahead of us. It would take us more than two million years to reach it (p. 136). In the course of our journey we would encounter hardly any stars and all around us in the black depths of space we would see only the faint light of extremely distant galaxies.

The Universe is a world of galaxies and galaxies are comprised first and foremost of stars. The total mass of a galaxy is between one thousand million and one billion solar masses. Besides stars, galaxies also contain thin gas, dust and solid bodies (such as planets and satellites). The size of galaxies ranges from several thousand to several hundred thousand light years.

Galaxies differ not only in size, but also in appearance, structure, age and evolution. The density of stars and interstellar gas and dust usually increases toward the centre (nucleus) of the galaxy. Galactic nuclei, however, are not mere clusters of stars; they are objects with quite extraordinary characteristics, with their own activity and source of energy, the origin of which is not reliably known as yet.

There are several tens of millions of galaxies within a radius of two to three thousand million light years from our planet. These join to form clusters sometimes numbering several thousand. Observations point to the fact that the distances between galaxies are continually increasing. We are part of an expanding Universe.

It is believed that the expansion of our Universe began with a big bang some 15 to 25 thousand million years ago. At a certain stage in the evolution of the Universe galaxies were formed from the matter dispersed by the explosion. The big bang must have been accompanied by radiation of extreme intensity which gradually decreased as the Universe expanded. The remnants of this radiation that fill the Universe (so-called relic or 3K radiation) were actually discovered in the mid-1960s, providing important evidence in support of the big bang theory.

Will the expansion of our Universe continue indefinitely? Or will this process one day come to a halt and reverse itself, with the Universe again contracting? For the time being we lack sufficient reliable data from observation to be able to decide which of the theoretically possible models of the Universe is closest to reality. New discoveries pose new questions. We may, however, assume with justification that even such difficult problems as the structure and evolution of the whole Universe are not unsolvable.

If science is described as an adventure of discovery, then this is unconditionally true of astronomy of the second half of the twentieth century. A new, more truthful picture of the Universe emerges before our eyes and it is up to each of us to make good use of it for the enrichment of our lives.

# FURTHER READING

Every observer will find *Norton's Star Atlas and Reference Handbook* (A. P. Norton, Gall & Inglis, Edinburgh, 16th edition 1973) indispensable as a guide book to the heavens.

Of the many books for the amateur by the prolific Patric Moore, two of the best for the beginner are
*The Amateur Astronomer* (Lutterworth, Guildford, 6th edition, 1967) and
*Naked Eye Astronomy* (Lutterworth, Guildford, 2nd edition, 1970) while for observers in the southern hemisphere his
*The Southern Stars* (Timmins, Cape Town; Rigby, Sydney; 1972) is equally good.

Amateurs interested in radio astronomy can find a great deal of practical advice in Frank Hyde's
*Radio Astronomy for Amateurs* (Lutterworth, Guildford, 1962) and more up-to-date information about the radio universe in J. S. Hey's *The Evolution of Radio Astronomy* (Paul Elek, London, 1973).

Some similar ground to that of the present volume is covered from a different viewpoint in John Gribbin's
*Astronomy for the Amateur* (Macmillan, London; McKay, New York; 1976), while the same author describes the modern picture of the Universe in *Our Changing Universe* (Macmillan, London; Dutton, New York; 1976).

At a slightly more technical level, and still non-mathematical, an excellent and up-to-date introduction to the whole subject of astronomy is provided by Robert Chapman's
*Discovering Astronomy* (Freeman, San Francisco, 1978).

For those interested in astronomical photography a lot of practical advice is available in Thomas W. Rackham's
*Astronomical Photography at the Telescope* (Faber and Faber, London, 1972).
Finally, two books which use the encyclopedic approach are worthy of mention:
*The Illustrated Encyclopedia of Astronomy and Space,* edited by Ian Ridpath (Macmillan, London; Crowell, New York; 1976) and
*Astronomy Data Book* by J. Hedley Robinson (Wiley, New York, 1972).

# INTRODUCTION TO THE PICTORIAL SECTION

The following 75 colour plates include three types of illustration. The first are views of celestial bodies, objects and phenomena as they appear to the observer on Earth. Further artists' impressions and diagrams are designed to give a more clear-cut idea of astronomical phenomena, and are presented from the viewpoint of a hypothetical extraterrestrial observer. The remaining plates consist of astronomical maps and diagrams.

In the plates showing astronomical objects North is at the top unless indicated otherwise. This applies also to the maps of the Moon, planets and starry sky, and should be kept in mind when making observations with a reversing telescope. In the section pertaining to the Solar System the usual symbols for the Sun, Moon and planets are used in the text accompanying the plates (Fig. 3).

The nomenclature in the astronomical maps is that which is authorized and used by the International Astronomical Union. The simplest nomenclature is to be found on the *star maps,* where the 88 basic constellations are designated by Latin names and individual stars and other objects by letters and numbers according to various catalogues. Only certain bright stars are still commonly designated by their names (including Vega, Sirius and Aldebaran).

Otherwise stars are usually designated by the small letters of the Greek alphabet (α, β, γ, etc.), generally according to diminishing brightness, or by so-called Flamsteed numbers (1, 2, 3, etc.), placed next to the genitive of the Latin name of the respective constellation. For example, the designation for the star Vega in the constellation Lyra is α (alpha) Lyrae. Star clusters and nebulae are generally identified with the numbers under which they are listed in Messier's catalogue, containing 103 bright objects (M 13, for example, is globular star cluster in Hercules), or in the New General Catalogue (NGC).

The nomenclature on the maps of the Moon and planets is more complex and contains hundreds of names, with new ones being constantly added. According to tradition, established by Riccioli in the mid-seventeenth century, lunar craters are named after prominent personalities, mostly scientists. There are more than 1,300 named craters on the Moon. Besides these, there are tens of other traditional names of the so-called seas, lakes, bays, and so on, as well as mountain ranges, valleys, and other features. The small scale of the *maps of the Moon* in this atlas makes it possible to identify only the most important formations by name. For better legibility and to save space, numbers and letters according

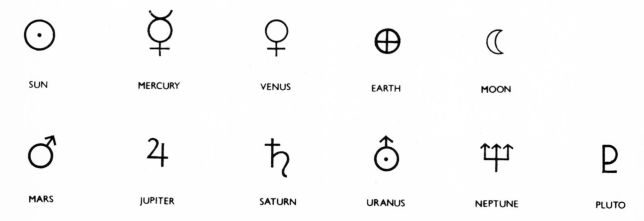

SUN     MERCURY     VENUS     EARTH     MOON

MARS     JUPITER     SATURN     URANUS     NEPTUNE     PLUTO

*Fig. 3. Astronomical symbols*

to the index on pp. 181—182 have been used on the maps to identify the various features.

The *maps of Mars* on pp. 81—91 are of two kinds: albedo maps (p. 81) and topographic maps (p. 89). The albedo maps show the dark and light areas on the planet's surface as they appear to the terrestrial observer. The nomenclature of the albedo formations dates from the nineteenth century. The topographic maps show the surface of Mars as photographed by space probes, mainly Mariner 9. The nomenclature of topographic features was determined by the International Astronomical Union according to the same principles as for the Moon. The craters on Mars are, like those of the Moon, named after prominent personalities.

The most recent maps are those of *Mercury* (see p. 75). Reliable albedo maps were not available until the late 1960s, when scientists finally succeeded in measuring the rotation period of Mercury (58.6 days) with the aid of radar. The beginnings of detailed mapping of the planet's surface are linked with the extremely successful photographs of Mercury made by the probe Mariner 10 which flew past the planet three times in the years 1974—1975. The sixteenth General Assembly of IAU in 1976 in Grenoble approved the nomenclature for the first map of the surface of Mercury, which included names for 7 plains (Planitia), 135 craters, one mountain range (Montes Caloris) and several valleys (Vallis), faults (Rupes) and ridges (Dorsum). It was decided to name the craters on Mercury after prominent artists — poets, writers, painters and composers. The one exception is the crater Kuiper, named after the astronomer who took part in the preparation of the Mariner 10 project. More than 60 per cent of Mercury's surface remains to be mapped.

# PICTORIAL SECTION

The novice amateur may find it useful to know how and when to observe the various phenomena and objects. To save space the following symbols are used in the marginal texts for this purpose:

*The individual parts of the Pictorial Section are marked with the following symbols:*

 Earth

 Moon

 Sun

 Planets

 Asteroids, comets, meteors

 Galaxies

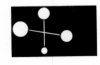 Star maps

 Stars, star clusters and nebulae

may be observed with the naked eye (visually)

may be observed with binoculars (visually)

may be observed with a telescope (visually)

may be photographed with a normal 35 mm or 2 ¼ camera

may be photographed with a telescope / camera combination

daily, may generally be observed every day

see astronomical year-book, may be observed only at a certain time or certain place of observation as stated in astronomical year-books (e. g. *Nautical Almanach*)

23

# DAY AND NIGHT
# TRANSMITTANCE
# OF THE ATMOSPHERE

Earth is our roving space laboratory. It moves through space and offers us an ever-changing panorama of the near and distant Universe. The plate on the opposite page shows the Earth as it appears in space. The blue band encircling the globe shows how the sky looks above the respective places on the Earth's surface. It is daytime on the side of the globe facing the Sun. The individual components (colours) of the Sun's light are scattered in the atmosphere unequally; the most greatly dispersed are short (blue) wavelengths, which is why the daytime sky is blue.

The atmosphere acts as a dense filter which absorbs a great part of the spectrum of electromagnetic radiation reaching it from space. The visible light and radio waves within the range of several millimetres to some 30 metres can only reach the surface of the Earth; there are just two 'windows' in the atmosphere, the optical and the radio. The other components of electromagnetic radiation are absorbed at various altitudes above the Earth's surface and can only be monitored from high-altitude balloons, rockets or space probes sent beyond the obscuring layers of the atmosphere.

$\lambda$ cm — wavelength in cm
G — gamma radiation
X — X-ray radiation
UV — ultraviolet radiation
V — visible light
IR — infrared radiation
R — radio radiation
the yellow arrows at right indicate the direction of the Sun's rays

1850 Photography first used in astronomy.
1932 Jansky (USA) — discovery of electromagnetic radiation of radio frequency from outside the Earth's atmosphere — beginnings of radio-astronomy.
1957 First artificail Earth satellite (USSR) — dawn of space astronomy.

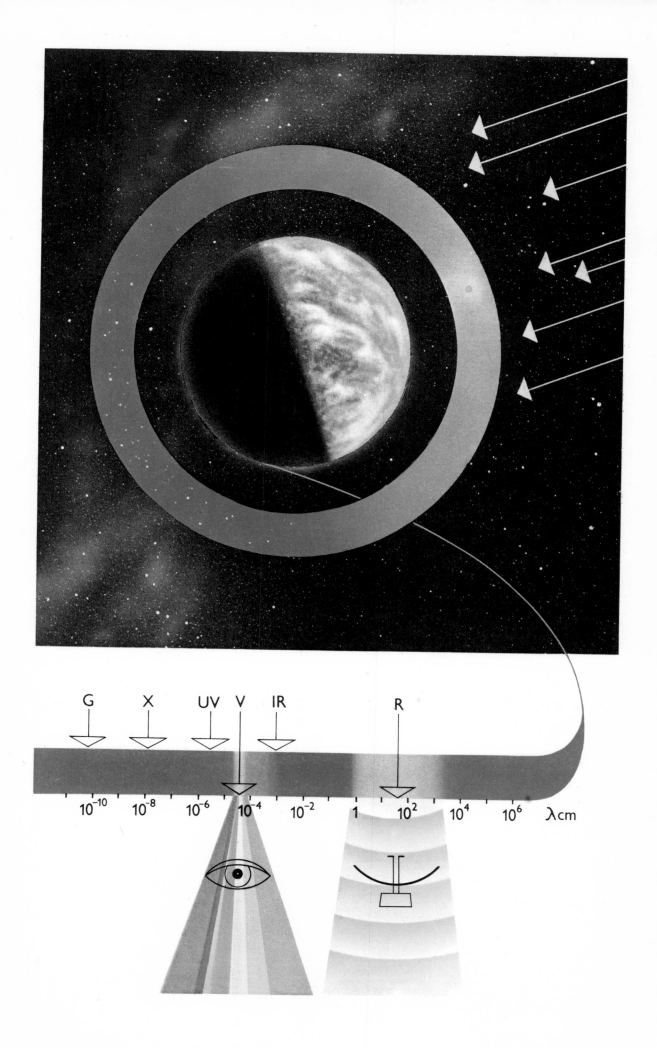

# LENGTH OF DAY
# AND NIGHT

Most astronomical observations are limited to the hours of darkness. The four diagrams show the changes in the length of the diurnal, twilight and nocturnal periods on the equator, at 40° and 50° geographical latitude, and at the North Pole. Twilight is defined in everyday life as the period after the Sun sets or before it rises when the Sun is less than 6° below the horizon (civil twilight); astronomers, however, extend the definition of astronomical twilight to the period when the Sun is 6° to 16° below the horizon. At high geographical latitudes (near the poles) twilight lasts till after midnight during the summer months.

Opposite, on the right of the diagram, is a schematic depiction of the Sun's orbit above the horizon at the times of equinoxes (when day and night are of equal length) and solstices (midsummer and midwinter) at the respective latitudes. On the equator the Sun's orbit is perpendicular to the horizon and the length of the period of daylight and darkness remains practically constant. Towards the pole the Sun's diurnal arc inclines towards the horizon and at the pole itself it is parallel to the horizon.

Example: At latitude 50° on 1 May night ends for the astronomer at 2 h 20 min and civil twilight begins at 4 h 00 min. The Sun rises at 4 h 35 min and sets at 19 h 20 min, civil twilight ends at 20 h 00 min and the ensuing night begins at 20 h 40 min.

h — hours,
d — date
a — vernal and autumnal equinox
b — summer solstice
c — winter solstice
A — night
B — astronomical twilight
C — civil twilight
D — day
N, S, E, W — compass points
O — observer

1582 *Introduction of the Gregorian calendar.*
1851 *Foucault (France) — demonstration of the Earth's rotation by means of a pendulum.*
1884 *The first 26 countries adopted standard time at a conference in the USA.*

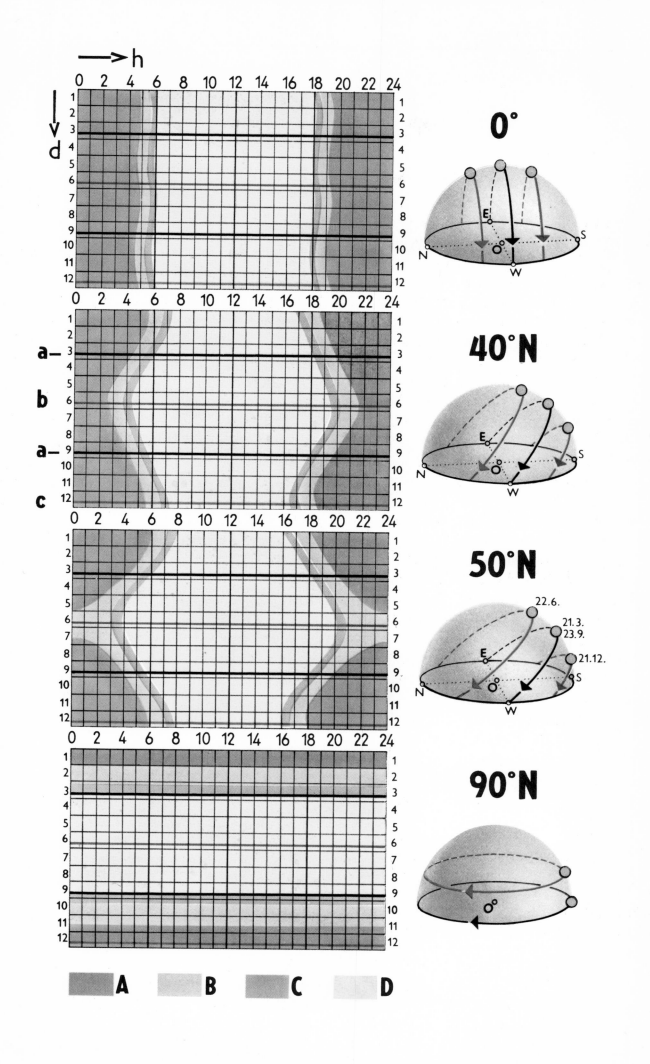

**0°**

**40°N**

**50°N**

**90°N**

22.6.
21.3.
23.9.
21.12.

A    B    C    D

# ATMOSPHERE OF THE EARTH
## INFLUENCE ON ASTRONOMICAL OBSERVATIONS

Light rays passing through the atmosphere to the Earth's surface change direction and are partially decomposed, dispersed and absorbed in the atmospheric layers. Let us imagine that the atmosphere is composed of thin layers increasing in density the closer they are to the Earth. A ray of light is always slightly refracted (bent) as it passes from one layer to another and reaches the Earth's surface from a seemingly different direction, from a seemingly greater height above the horizon. The resultant deviation, known as *astronomical refraction*, is greatest at the horizon, where it equals approximately 35′. When the Sun or Moon (which each has an angular diameter of about 30′) appear to be touching the horizon with their lower limb, they are in reality already below the horizon. Because the refraction is greater in the lower limb, the Sun (Moon) has flattened appearance near the horizon.*)

The unevenness, temperature differences and motion of the atmospheric layers lead to complex deformations of the solar disc near the horizon. The wave motion of the atmospheric layers and the brief, rapid changes in the direction of the light rays markedly influence the quality of the conditions of observation (seeing). In the case of atmospheric turbulence the image in the telescope 'boils' and is blurred and lacking in detail.

---

*) In angular measure each degree is divided into 60 minutes of arc, 60′, and each minute into 60 seconds of arc, 60″ . 30′ is the same as half a degree.

Deformation of the Sun or Moon near the horizon:

daily

S, $S_1$ — real position of the Sun
S′, $S_1'$ — apparent position of the Sun
O — place of observation
h — horizon
a — astronomical refraction
b — deformation of the Sun near the horizon
c — turbulence in the atmosphere and 'seeing'

*1609—1610 Galileo Galilei (Italy) — first astronomical observations with a telescope.*
*1668 Newton (England) — first reflecting telescope.*
*1758 Dollond (England) — invention of the achromatic lens.*

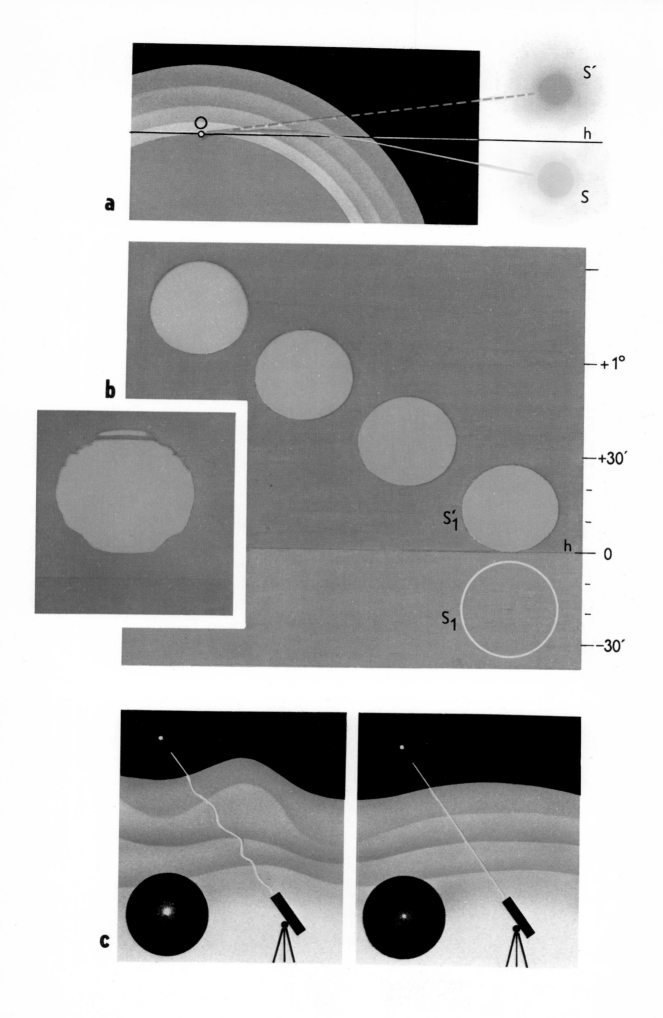

a

b

S′₁

S₁

+1°

+30′

h 0

−30′

c

# EARTH'S MAGNETOSPHERE

Earth's magnetic field creates a vast drop-shaped space round the planet, known as the magnetosphere, encircled by a stream of solar plasma (gas in which atoms are broken down into separate particles, positively charged nuclei and negative electrons) composed of ionized particles (solar wind). This magnetosphere is also filled with plasma. Periods of increased solar activity are often accompanied by changes in the magnetosphere giving rise to such phenomena as magnetic storms and auroras.

Auroras are a visible form produced by plasma radiation originating in the magnetosphere and spilling out along the lines of force of Earth's magnetic field into the upper layers of the atmosphere at altitudes of approximately 100 to 1,000 kilometres. These optical phenomena are common in the polar regions; on rare occasions they may be observed as far away as the tropics. Their forms, motion and colours are extremely varied and beautiful. In substance they are divided into two groups: those with a uniform (homogeneous) structure (such as a homogeneous quiet arc, b) and those with a ray-structure (arc with ray-structure, d; draperies, c; and corona, e).

 (high speed colour film)

chiefly during periods of maximum solar activity

a — schema of the Earth's magnetosphere
1 — solar wind
2 — shock wave
3 — magnetopause
4 — radiation bands
b, c, d, e — forms of auroras

*1958—1960 Van Allen (USA), Vernov et al. (USSR) — discovery of radiation belts round the Earth; beginnings of the study of the Earth's magnetosphere by means of artificial satellites.*

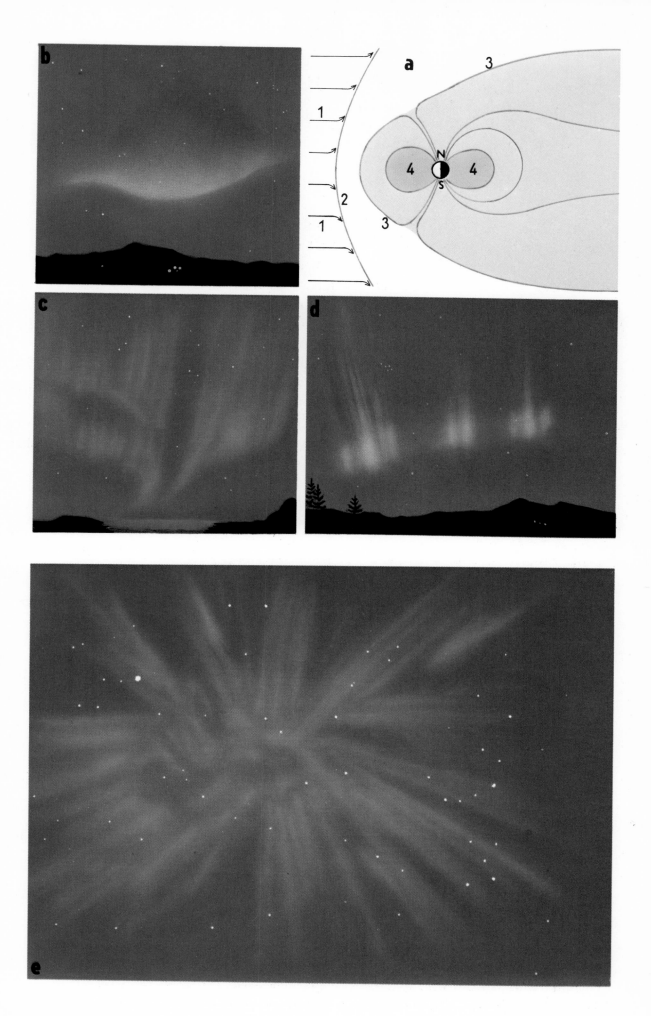

The Moon is a lifeless 'mini-planet' (3,476 km in diameter) which revolves round the Earth at a mean distance of 384,000 kilometres and appears in the sky as a disc 0.5° (30′) in diameter. At any moment half of the Moon is illuminated by the Sun, the other is in its own shadow; the line dividing the two is called the *terminator.* From Earth we can see only part of the illuminated hemisphere. That is why we see the Moon in various phases.

When the Moon is betweeen the Sun and Earth (position 1), the daylit side cannot be observed — this is the phase of New Moon. The *Moon's 'age'* is counted in days starting from this point. The first few days after the New Moon it appears as a crescent with the remainder of the disc faintly illuminated by Earthshine (sunlight reflected from the Earth's surface). When it is 7.4 days old, the Moon is in its first quarter (2). The Full Moon (3) is when the entire nearside hemisphere is illuminated and it is seen as a full disc. At the age of 22.1 days the Moon is in its third quarter (4) and after 29.51 days, i.e. after one *synodic month,* it is back again to the New Moon and the next lunar month starts.

At the terminator the angle of incidence of the Sun's rays is very small and the shadows cast on the Moon's surface highlight the details of the lunar landscape, craters, mountain ranges, etc.

daily except for the New Moon

S — direction of the Sun's rays
1, 5 — New Moon
2 — first quarter
3 — Full Moon
4 — third quarter
d — days (Moon's age)
a — Earth's orbit
b — Moon's orbit

*433 B.C. Meton (Greece) — discovery of the 19-year cycle: during the cycle the phases of the Moon occur on the same dates as 19 years previously.*

*1500 Leonardo da Vinci explains the cause of Earthshine.*

*1923 Commencement of the numbering of lunar months from New Moon to New Moon (e.g. 26 February 1979 marks the beginning of lunar month — lunation no. 695).*

# ECLIPSE
# OF THE MOON

When the Full Moon passes close to the imaginary line connecting the centre of the Sun and the centre of the Earth, it may move into the shadow cast by the Earth and thus be eclipsed. The Earth casts a cone of complete shadow (umbra, U) surrounded by a lesser shadow (penumbra, P) into space. If the Moon is in the penumbra (1), we speak of a penumbral lunar eclipse marked by only a slight decrease in the Moon's brightness. If the whole Moon is obscured (2), we speak of a *total eclipse;* if only part of the Moon is obscured (3), we speak of a *partial eclipse.*

As they pass through the atmosphere A, the Sun's light rays are bent (refraction) and enter the Earth's geometric shadow. The apex of the umbral shadow cone X is therefore closer to the Earth than the Moon, which is faintly illuminated even during a total eclipse. However, it is primarily the red rays *(r)* that pass through the atmosphere into the Earth's shadow, for the other spectral components of sunlight are dispersed. The result is that the Moon is coloured red, or reddish-brown, during a total eclipse.

Over the whole twentieth century there will be a total of 148 lunar eclipses. Unlike an eclipse of the Sun, a lunar eclipse may be observed from any point on that half of the Earth where the Moon is above the horizon at the time.

rare phenomenon — see astronomical year-book and the tables on pp. 160—174

a — schema of a lunar eclipse
b — course of the eclipse
c — partial lunar eclipse
S — direction of the Sun's rays
r — red edge of the spectrum
v — violet edge of the spectrum

*3379 B.C. Ancient Mayan record of the eclipse of the Moon.*
*1887 Oppolzer (Austria) — tables of solar and lunar eclipses from the year 208 B.C. to 2163 A.D. (they contain 8,000 solar eclipses and 5,200 lunar eclipses).*

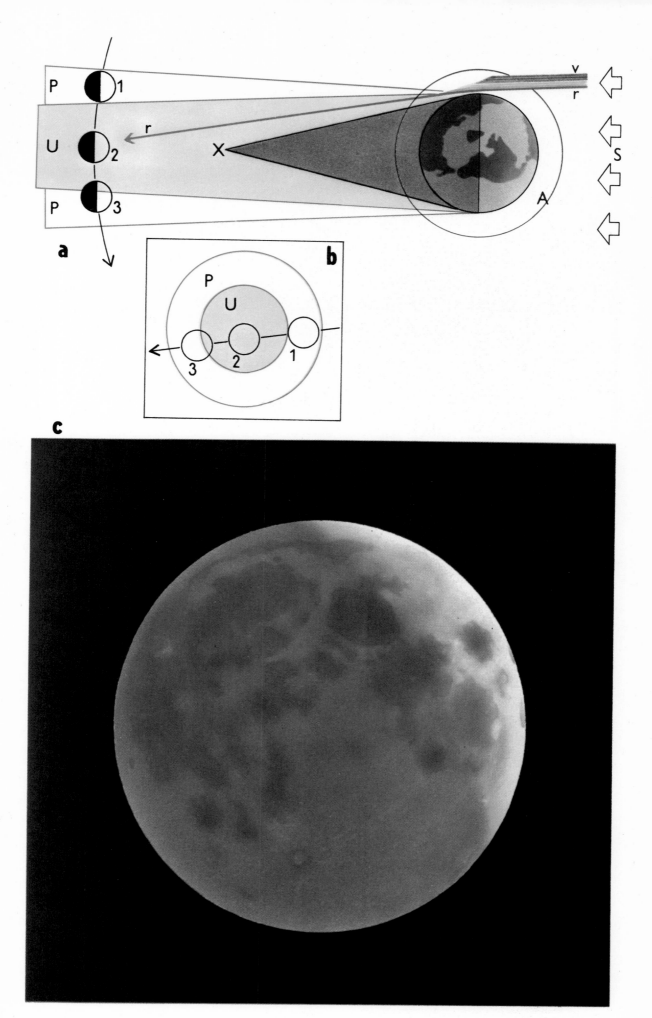

a

b

c

# ECLIPSE
# OF THE SUN

When the New Moon passes close to the imaginary line connecting the centre of the Earth and the centre of the Sun the Moon's penumbra (2) and umbra (1) may cover a certain part of the Earth, from which it is then possible to observe the solar eclipse.

To an observer standing in the penumbra of the Moon part of the Sun is obscured by the lunar disc and he sees a *partial solar eclipse.* The penumbra of the Moon covers an area on the Earth measuring several thousand kilometres in diameter, so that a partial eclipse of the Sun may always be observed over that wide area. On the other hand, the area covered by the Moon's umbra never exceeds 270 kilometres in diameter. So a total solar eclipse is a much rarer event at any point on Earth. The rotation of the Earth and revolution of the Moon cause the shadow cast on the Earth to move at a speed of more than 2,000 km/h, so that a *total eclipse* of the Sun lasts a few minutes (7.6 minutes at the most) and may be observed only from a narrow *band of totality* (3). During a total eclipse the observer can also see with the unaided eye bright stars and, round the black disc of the Moon, the pinkish glow of the solar chromosphere with prominences and the silvery corona. It is one of the most beautiful and rarest of natural phenomena.

The Moon's orbit around the Earth is an elliptical one, sometimes taking it further away from us than at other times. If it is at a far point during an eclipse, then it appears to be smaller than the Sun and its umbral shadow does not reach the surface of the Earth; the resulting phenomenon is called an *annular eclipse.*

Over the whole of the 20th century there will be 228 total and partial eclipses.

rare phenomenon — see astronomical year-book and the tables on pp. 160—174

S — Sun
M — Moon
1 — umbra
2 — penumbra
3 — band of totality
a — total eclipse of the Sun
b — partial eclipse
c — annular eclipse

*1860 Secchi (Italy) and De La Rue (England) — first to succeed in photographing the corona and prominences during a solar eclipse. Proof that the corona is a part of the Sun.*
*1931 Lyot (France) — first photographed the Sun's corona at a time other than during an eclipse. Invention of the coronagraph.*

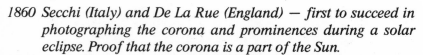

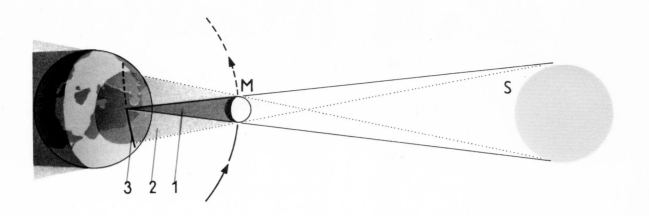

a

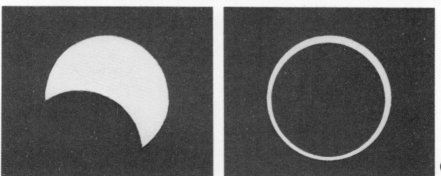

b

c

Even without a telescope one can see the dark areas on the Moon called *seas* (*mare* in Latin). The light areas are called continents. The forms and outlines of the 'seas' are useful for orientation on the Moon and it is helpful to know their names (for translation of Latin names see p. 182).

The seas do not contain water, but are solidified lava covers that fill the vast basins created by the impact and explosion of large meteoritic bodies. The surface of the seas is comparatively smooth and a small telescope reveals few details. However, with the aid of a telescope the lighter areas are revealed as craters, mountain ranges, hills, and the like (see p. 43). Full Moon is the best time to see the bright rays extending outwards from the craters Tycho, Copernicus, Kepler, Aristarchus, etc. But other lunar features are hard to distinguish at this time, when the angle of incidence of the Sun's rays is large; visibility is far better in the oblique illumination near the terminator (the line dividing the illuminated and dark parts of the disc) at times other than at Full Moon.

For more detailed and graphic information on the near side of the Moon see maps L 1 — L 6, pp. 47—57.

daily except at Full Moon

M. C. — Mare Cognitum
M. V. — Mare Vaporum
S. A. — Sinus Aestuum
S. I. — Sinus Iridum
S. M. — Sinus Medii
S. R. — Sinus Roris
L 1 — L 6 — maps of the Moon on pp. 47—57

*1817 Frauenhofer (Germany) — the Moon and planets have a solar spectrum — proof that they shine by reflected light.*
*1839 Daguerre (France) — first attempt at photographing the Moon.*

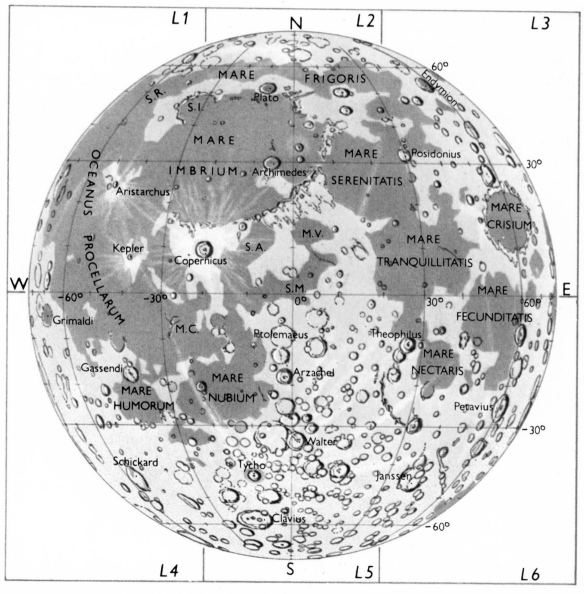

L1     N     L2     L3

60°

S.R.    MARE    FRIGORIS    Endymion

Plato

S.I.

MARE

MARE    Posidonius    30°

O C E A N U S    I M B R I U M    Archimedes    SERENITATIS

Aristarchus

MARE
CRISIUM

Kepler    M.V.    MARE

S.A.    TRANQUILLITATIS

Copernicus

P R O C E L L A R U M

W    −60°    −30°    S.M.    0°    30°    60°    E

MARE

Grimaldi    FECUNDITATIS

M.C.    Ptolemaeus    Theophilus

Gassendi    MARE
NECTARIS

MARE    MARE    Arzachel

HUMORUM    NUBIUM    Petavius

−30°

Walter

Schickard    Tycho    Janssen

Clavius    −60°

L4     S     L5     L6

# FAR SIDE
# OF THE MOON
# BOUND ROTATION

The Moon always presents the same face to the Earth, which is sometimes mistakenly believed to indicate that the Moon does not rotate round its axis. If it truly did not rotate, then its motion round the Earth would be as shown in Figure a. In fact the Moon rotates round its axis within exactly the same period as it revolves round the Earth (bound rotation — Figure b). For example, from position 1 b to position 3 b the Moon travels one-quarter of its orbit and simultaneously rotates by one-fourth (i.e. 90°) round its axis. One hemisphere (hatched in Figure b) thus remains permanently turned away from the Earth but the entire lunar sphere is successively illuminated by the Sun (yellow hatching).

The far side of the Moon has now been mapped in detail by space probes. On the far side there are several small 'seas' covered with dark material. Light areas prevail; these have a greatly diversified topography and are covered with craters like the continents on the side facing the Earth.

A slight oscillation of the Moon in the sky (so-called libration) makes it possible for terrestrial observers to gradually see near the limb of the Moon approximately nine per cent of the far surface.

limb of the Moon under favourable libration — see astronomical year-book

Mare Moscoviense — Moscow Sea
Mare Ingenii—Sea of Ingenuity
Mare Orientale — Eastern Sea

*1959 4 October — Luna 3 space probe (USSR) — first photographs of the far side of the Moon.*
*1967 Far side of the Moon first mapped in detail in the USA on the basis of photographs made by the lunar Orbiter probes.*

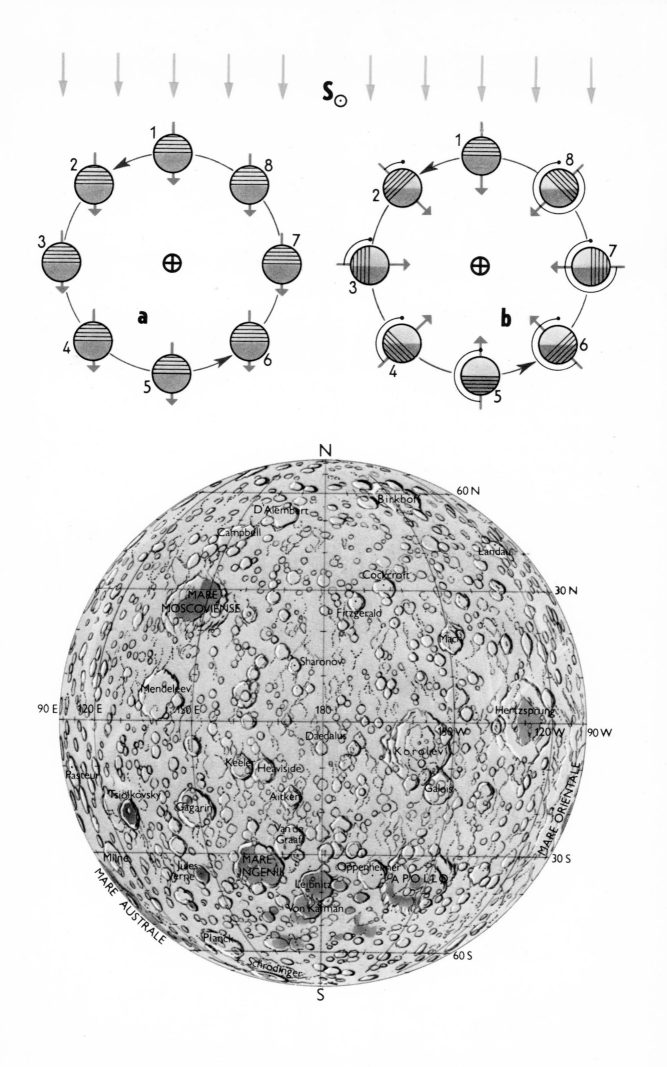

# SURFACE
# OF THE MOON
# TYPICAL
# FORMATIONS

The Moon is a very rewarding object for observation by the amateur; many interesting details may be seen even with a small telescope. Typical formations on the Moon (and on other objects in the solar system) are craters of varying dimensions. The largest are the *walled plains,* such as Ptolemaeus (153 kilometres in diameter and 2,400 metres deep). *Ring mountains,* such as Copernicus (diameter 93 kilometres; depth 3,760 metres), have terraced walls of varied topography. The lunar Alps (Montes Alpes) are an example of a typical *mountain range,* the various peaks reaching heights of about 2,000 metres above the level of the adjacent Mare Imbrium.

Under conditions of oblique illumination the long, black shadows make the lunar mountains seem very steep with pointed peaks; likewise, the craters look like deep holes with steep walls. In reality the gradient of the slopes on the Moon only rarely exceeds 30°.

In the lunar Alps there is a 130 kilometre long valley called Vallis Alpes (top right in Figure b). With a larger telescope it is possible to observe *rilles* and *clefts* in many places. A striking example is the rille by the crater Hyginus. The system of clefts by the crater Triesnecker is a suitable object for testing the quality of a telescope.

The depicted details correspond to observations made with a roughly 15 cm telescope.

Details of the lunar surface:

in all phases, particularly near the terminator

a — walled plain
b — mountain range, mountains
c — ring mountain
d — rilles, clefts

*1830s Beer and Mädler (Germany) — detailed map of the Moon and proof that the Moon is a celestial body without an atmosphere and without water.*
*1927 Pettit and Nicholson (USA) — measurements of temperature fluctuations on the Moon.*

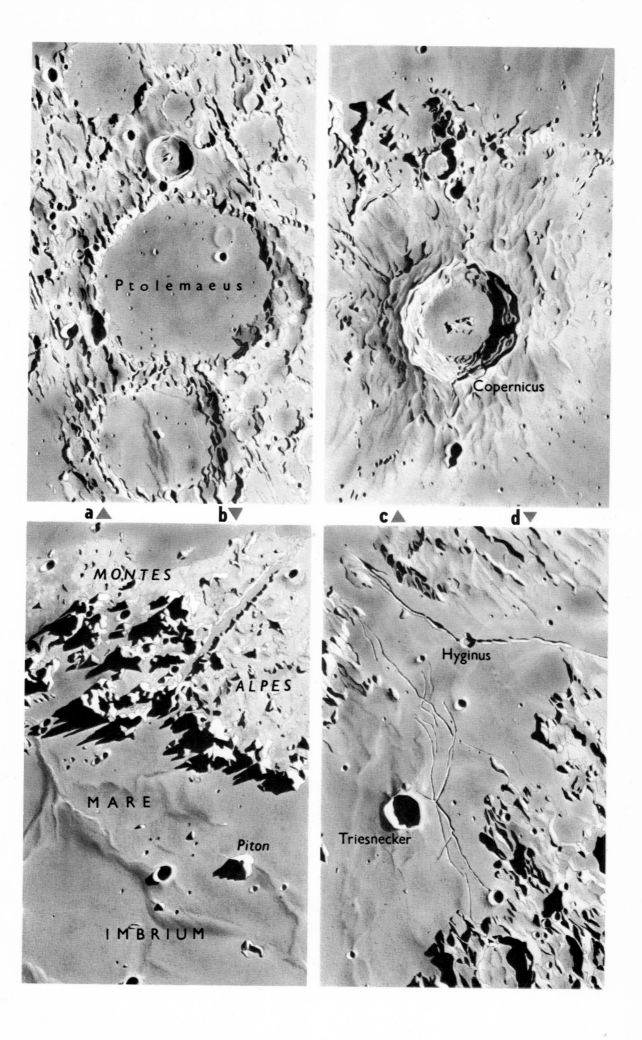

Ptolemaeus

Copernicus

a ▲          b ▼          c ▲          d ▼

MONTES

ALPES

Hyginus

MARE

Piton

Triesnecker

IMBRIUM

# MAP
## OF THE NEAR SIDE
## OF THE MOON
# LAYOUT OF MAPS
# L 1 — L 6

Knowing the names of the seas, mountain ranges and more important craters is a great aid to quick orientation on the Moon. Exact and explicit positions of various points on the Moon are given by *selenographic coordinates:* selenographic longitude λ (similar to the geographic longitude) and selenographic latitude β (similar to the geographic latitude). Selenographic longitude is the angle between the central meridian and the meridian containing the given point. It is measured positively to the east and negatively to the west, always from 0° to 180°. Selenographic latitude is the angular distance of the given point from the equator, reckoned on the given meridian positively to the north and negatively to the south, always from 0° to 90°.

Conditions for observing the various parts of the Moon's surface are determined not only by the angle of incidence of the Sun's rays, but also by the Moon's position towards the Earth.

The angle of incidence of the Sun's rays and position of the terminator can be roughly estimated according to the age of the Moon *d* (see p. 32), which can be easily told, for example, from the tables giving the phases of the Moon for the years 1979—2000 on pp. 160—174. It must be kept in mind, however, that the position of the terminator in relation to the lunar formations is not always the same for the given age of the Moon. As a result of the Moon's uneven orbital velocity, the inclination of its rotational axis to the ecliptic plane and other reasons, the Moon oscillates slightly in all directions towards the Earth. This oscillation is called libration and causes marked changes in the appearance of formations on the limbs of the Moon (Mare Crisium, for example, changes from a narrow ellipse to a broad oval, and so on) and apparent shifts of all points on the Moon in relation to the centre of the lunar disc. Thus it happens that sometimes the crater Archimedes and the Alps and Apennines cannot be seen in the first quarter (age 7.4 days), whereas at other times in the same lunar phase when libration is more favourable these formations are flooded with the light of the rising Sun.

If we wish to know beforehand at what areas the Sun will rise and set, it is necessary to look up the *co-longitude,* which is the selenographic longitude of the morning terminator, in the astronomical yearbook.

*1651 Riccioli (Italy) — foundations of the current nomenclature for maps of the Moon.*

*1935 Blagg (England) and Müller (Germany) — international nomenclature for the Moon submitted by the two scientists is accepted by the International Astronomical Union.*

daily except for the period of the New Moon

d — age of the Moon in days
L 1 — L 6 — numbers of the maps of the Moon on pp. 47—57.

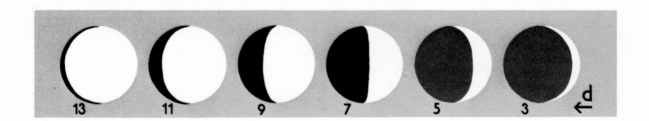

13　11　9　7　5　3　←d

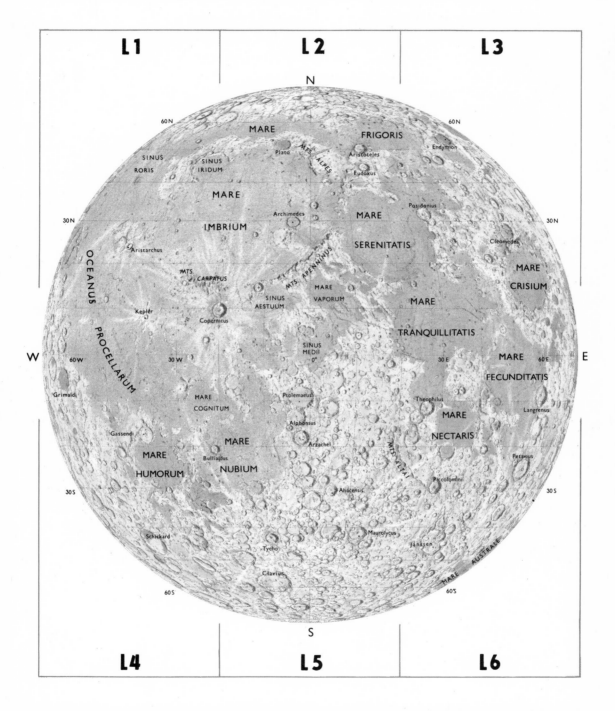

**L1**　　**L2**　　**L3**

N

60N

MARE
FRIGORIS

SINUS
RORIS

SINUS
IRIDUM

Plato

MTS. ALPES

Aristoteles

Endymion

60N

Eudoxus

MARE

Posidonius

MARE

IMBRIUM

Archimedes

SERENITATIS

Cleomedes

30N

30N

Aristarchus

MARE

CRISIUM

MTS. CARPATUS

MTS. APENNINUS

MARE
VAPORUM

MARE

OCEANUS

Kepler

SINUS
AESTUUM

Copernicus

TRANQUILLITATIS

PROCELLARUM

SINUS
MEDII
0°

MARE
FECUNDITATIS

W

60W

30W

30E

30E

60E

E

Grimaldi

Ptolemaeus

Theophilus

Langrenus

MARE
COGNITUM

Alphonsus

MARE

Gassendi

Arzachel

NECTARIS

Petavius

MARE

Bulliaidus

MARE

Piccolomini

HUMORUM

NUBIUM

MTS. ALTAI

30S

Altacensis

30S

Schickard

Maurolycus

Janssen

MARE AUSTRALE

Tycho

60S

Clavius

60S

S

**L4**　　**L5**　　**L6**

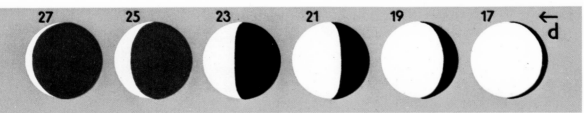

27　25　23　21　19　17　←d

Map L 1 shows the northwestern part of the near side of the Moon dominated by the dark areas of the lunar seas Oceanus Procellarum (O.P.) and Mare Imbrium (Im). Clearly visible at Full Moon are the bright rays extending outwards from the craters Copernicus (No. 90 on the map; diameter 93 km), Kepler (166; 32 km) and Aristarchus (23; 45 km). Aristarchus is one of the brightest formations on the Moon; it is visible even in Earthshine (the reflected light from Earth faintly illuminating the dark part of the Moon). Near Aristarchus is the sinuous valley Vallis Schröteri (v; length of valley 200 km, depth 1,000 m, width up to 10 km), visible even with a small telescope. Distinctive features at the edge of Mare Imbrium are the crater formation Sinus Iridum (S.I.; 260 km) and the lunar Carpathian Mountains (c).

Southwest of the crater Copernicus lies the small crater Hortensius (158; 14.6 km) and nearby a group of typical lunar domes (marked by crosses on the map); they are rounded, elevated formations 10 to 20 kilometres in diameter and only several hundred metres high, visible only close to the terminator. West of Sinus Iridum is the elevated formation Rümker (s; 55 km), a solitary complex of domes.

North of the crater Marius (191; 41 km) extends a typical rille approximately 2 kilometres wide and 250 kilometres long.

age of the Moon 9 d — 15 d, 22 d — 28 d

name index see p. 181 ff.
x domes
— · — · — rilles

*1922—1927 Lyot (France) — on the basis of the study of the polarization of moonlight he deduced that the Moon is covered with a layer of dust.*
*1966 3 February — Luna 9 space probe (USSR) — first soft landing and proof that the dust layer is thin and that the lunar surface is firm enough for machines and man to land there.*

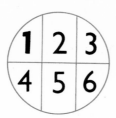

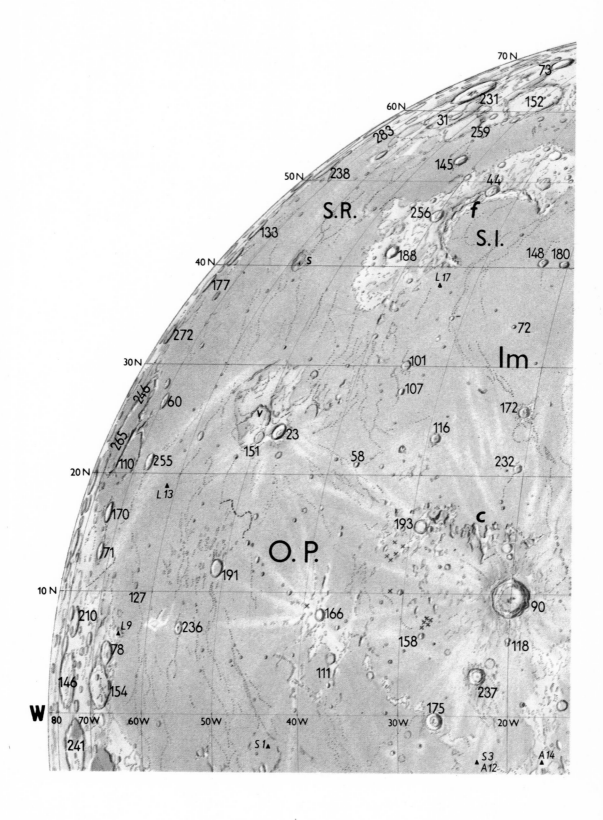

70 N

73

60 N · 231 · 152

31 · 259

283 · 145

50 N · 238 · 44

S.R. · f

256 · S.I.

133 · 148 · 180

40 N · s · 188

L 17 · 72

177

Im

272 · 101

30 N · 107

246 · 60 · 172

116

265 · 23 · 58 · 232

151

110 · 255

20 N

L 13

170 · c

193

71 · O. P.

191

10 N · 90

127

210 · 166

L 9 · 236 · 158 · 118

78

111 · 237

146 · 175

154

W · 80 · 70 W · 60 W · 50 W · 40 W · 30 W · 20 W

241

S 1 · S 3 · A 14
A 12

Central and northern part of the near side of the Moon.

Mare Imbrium (Im) is the largest of the lunar basins formed by the vast impact and then gradually filled in with lava. The basin is rimmed by a chain of mountain ranges: Montes Alpes (a), Montes Caucasus (d), Montes Apenninus (b), Montes Carpatus (c) and Montes Jura (f, map L 1). These mountain ranges are the remnants of the wall of what was once a giant crater measuring 1,250 kilometres in diameter. The peaks of the Apennines rise to 5,000 metres above the level of Mare Imbrium.

The region of the distinctive trio of craters Archimedes (No. 20 on the map; diameter 83 km), Autolycus (29; 39 km) and Aristillus (24; 55 km) is where the Soviet space probe Luna 2 (L 2) — first of mankind's messengers — crash-landed in 1959. The floor of the crater Plato (223; 100 km) is one of the darkest areas on the Moon.

Located near the centre of the near side (at bottom of map) are the craters Triesnecker (270; 26 km) and Hyginus (160; 10.6 km) and a well-known complex of valleys and clefts (see p. 43).

The crater Bessel (42; 16 km) in Mare Serenitatis (Se) is intersected by the longest of the bright rays extending from the crater Tycho (see section L 5). The small crater Linné (183; 2.1 km) is surrounded by light material which, when the Sun is high in the lunar sky (near Full Moon), looks like a bright patch.

age of Moon — 6 d — 12 d,
19 d — 24 d

name index — see p. 181 ff.
— · — · —  rilles

*1968 Muller and Sjogren (USA) — analysis of observations made by the lunar Orbiter spacecraft, discovery of 'mascons' — concentrations of matter in Mare Imbrium and other seas.*
*1971 Apollo 15 (USA) — wheeled vehicle first used by astronauts for travel on the Moon.*

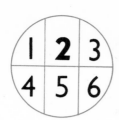

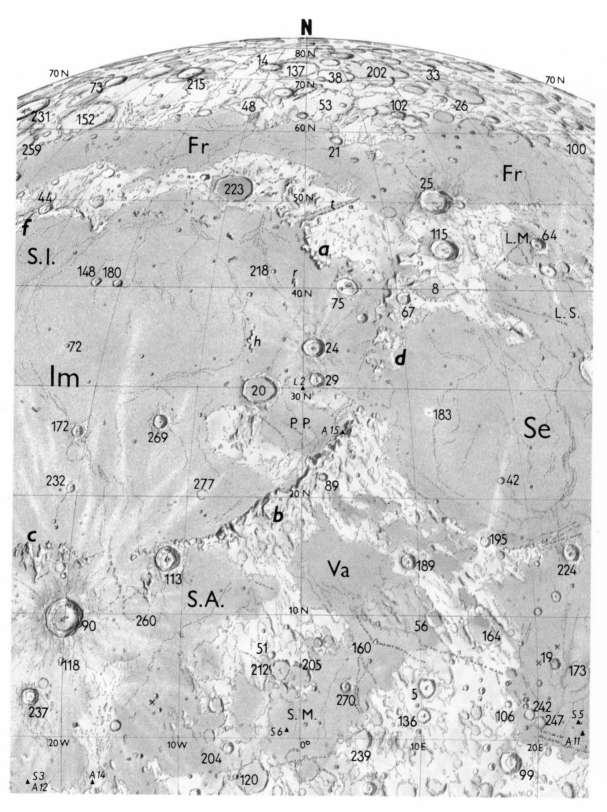

N

70 N
80 N
70 N
60 N
50 N
40 N
30 N
20 N
10 N
0°

73
14
137
38
202
33
215
48
53
102
26
231
152
21
259
Fr
Fr
223
25
44
t
f
a
115
L.M.
64
S.I.
218
r
8
148  180
75
67
L.S.
72
h
24
d
Im
20
L 2  29
183
Se
172
P.P.
269
A 15
42
232
277
89
b
c
195
189
224
Va
113
S.A.
56
90
260
164
19
173
51
160
118
212
205
5
242
237
270
136
106
247
S 5
S.M.
99
S 6
A 11
204
239
120

20 W
10 W
0°
10 E
20 E

S 3
A 14
A 12

Northeastern part of the near side of the Moon.

Several unmanned lunar probes and two Apollo missions landed in this region. The southwestern edge of Mare Tranquillitatis (Tr) is where man first set foot on the Moon (see table on p. 183). In the eastern part of this sea is the small crater Cauchy (No. 77 on the map; diameter 12.4 km), next to which, under oblique illumination, the cleft and fault Rupes Cauchy (n) stands out clearly.

The dark elliptical Mare Crisium (Cr) may be seen clearly even without a telescope. During favourable libration (turning of the Moon towards the Earth) it is possible to observe on the Moon's eastern limb the greatly distorted shapes of Mare Marginis (Ma) and Mare Smythii (Sm — see also section L 6). On the Moon's northeastern limb one can see the small Mare Humboldtianum (Hm) and nearby the dark-floored crater Endymion (112; 125 km).

The crater Hercules (150; 67 km) and Atlas (28; 87 km) form a striking twosome. On the floor of the walled plain Posidonius (227; 100 km) numerous clefts may be observed with a larger telescope. Countless marial wrinkle ridges, resembling prominent veins, may be seen, for instance, in the vicinity of the formation Lamont (173; 88 km); these may be observed in the other lunar seas as well.

age of Moon — 1 d — 7 d,
15 d — 21 d

name index see p. 181 ff.
x dome
— · — · — rille
⌐┬┬┬┬┬┐ fault

*1969 20 July — Expedition Apollo 11 (USA) — first men on the Moon.*
*1970 21 September — Luna 16 space probe (USSR) — first automatic collection of a lunar rock sample brought back to Earth.*

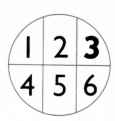

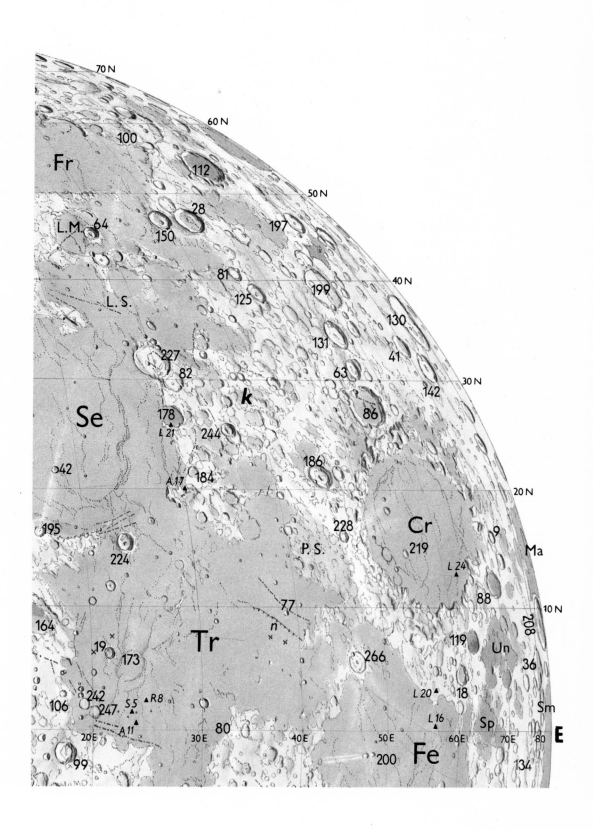

70 N

60 N

100

Fr

112

50 N

28

L.M. 64

197

150

81

L.S.

125

199

40 N

130

131

227

41

82

63

30 N

k

142

Se

86

178

L 21

244

186

42

184

A 17

20 N

228

Cr

195

P. S.

219

9

Ma

224

L 24

88

77

208

10 N

n

164

119

Un

36

19

Tr

173

266

18

L 20

242

L 16

Sm

106

247

S 5

R 8

Sp

20 E

A 11

30 E

80

40 E

50 E

60 E

70 E

80

E

99

200

Fe

134

The largest of the lunar seas — Oceanus Procellarum (O.P.) extends from the north into the southwestern part of the near side of the Moon. On the northern edge of Mare Humorum (Hu) lies the walled plain Gassendi (No. 128 on the map; diameter 110 km). A distinctive feature in Mare Nubium (Nu) is the ring mountain Bullialdus (62; 59 km). In the region of Palus Epidemiarum (P.E.) and on the eastern edge of Mare Humorum there are numerous rilles.

Mare Cognitum (Co) — Known Sea — was named after the successful flight of Ranger 7, which before landing (at point R 7) transmitted to Earth the first detailed photographs of the lunar surface with detail of features down to a size of approximately one metre.

Adjoining the walled plain Schickard (252; 227 km) is the remarkable crater Wargentin (279; 84 km), filled almost to the rim with lava; there are very few such formations on the Moon. The double crater Schiller (253; 179 km × 71 km) is shaped like a footprint.

During favourable librations it is possible to see on the SSW limb of the Moon one of the largest of the walled plains Bailly (34; 303 km). Visible on the western limb are the lunar Cordilleras (e) and other mountain ranges in concentric circles bordering the basin of Mare Orientale, the centre of which is located on the far side of the Moon.

age of Moon 9 d — 15 d,
22 d — 28 d

name index see p. 181 ff.
x dome
— · — · — rille, cleft

*1964 31 July — Ranger 7 space probe (USA) — first close-ups of the Moon showing features smaller than one metre, i.e. resolution 1,000 times greater than possible with the best telescopes from Earth.*

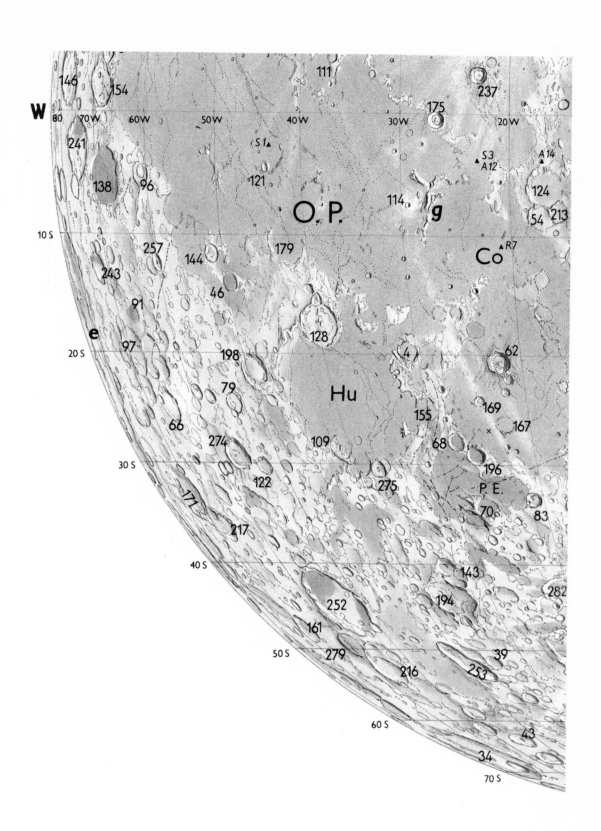

W

80 70W 60W 50W 40W 30W 20W

146
154
111
237
175
241
S 1
S 3
A 12
A 14
138 96
121
O. P.
114
g
124
54 213
10 S
257
144
179
Co
R7
243
46
91
128
e
97
198
62
20 S
79
4
66
Hu
155
169
274
109
167
122
68
171
275
196
217
70
P. E.
83
143
40 S
282
252
194
161
279
39
216
253
50 S
43
60 S
34
70 S

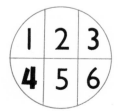

| 1 | 2 | 3 |
|---|---|---|
| **4** | 5 | 6 |

**L 4**

The central and southern part of the Moon consists mostly of a light 'continent', dotted with numerous craters. The chemical composition of the light continents is rich in calcium and aluminium; the dark seas contain more titanium, iron and silicon.

In Mare Nubium (Nu), between the craters Thebit (No. 267 on the map; diameter 55 km) and Birt (49; 17 km) the best-known of the lunar faults — Rupes Recta (p) is located. Its length is 96 kilometres, height up to 300 metres, apparent width 2.5 kilometres and gradient of slope only 7°, even though under oblique illumination it appears to be a steep slope.

The crater field in the southern hemisphere is dominated by the circular mountain range Tycho (271; 85 km), centre of the most striking bright rays. The depth of this crater is 4,850 metres and from its floor rises a central massif 1,600 metres high.

For purposes of general orientation in this region it is useful to remember certain groups of large craters such as:

Clavius (85; 225 km), Maginus (187; 163 km), Longomontanus (185; 145 km)

Ptolemaeus (229; 153 km), Alphonsus (13; 118 km), Arzachel (27; 97 km); see also detail on p. 43.

Stöfler (264; 137 km), Maurolycus (192; 114 km)

age of Moon 6 d — 12 d,
19 d — 24 d

name index see p. 181 ff.
⌐ⁱⁱⁱⁱⁱⁱⁱ fault

*1958 Kozyrev (USSR) — observation of an emission of gas in the crater Alphonsus.*

*1965 24 March — Ranger 9 space probe (USA) — detailed pictures of the floor of the crater Alphonsus with resolution of up to 0.5 metres.*

S. M.

S 6 ▲

237
20 W
10 W
0°
270
5
136
106
242
247
S 5 ▲
A 11 ▲
204
239
10 E
20 E
99
S 3 ▲
A 12 ▲
A 14 ▲
120
156
285
124
153
A 16 ▲
104
54
213
229
10 S
7
15
2
94
Co
R 7 ▲
139
98
R 9 ▲
13
22
11
76
12
6
169
62
108
1
30
225
49
p
267
248
m
Nu
230
17
167
234
281
196
221
10
P. E.
105
278
132
233
70
83
129
61
65
147
264
192
143
S 7 ▲
211
207
37
282
250
181
84
222
271
149
39
185
187
93
32
276
253
182
162
157
245
85
284
206
141
43
251
50
190
34
168
203
92
52
57
74
70 S
70 S
254
103
S
80 S

L 5

1 2 3
4 **5** 6

Southeastern part of the near side of the Moon.

In Mare Fecunditatis (Fe) there is a prominent pair of small craters: Messier (No. 200 on the map; diameter 10 km) and Messier A (west of Messier), from which two bright rays extend farther west.

The basin of Mare Nectaris (Ne) is bordered by the Altai fault — Rupes Altai (m) — like part of a circular wall, terminating in the south in the circular mountain range Piccolomini (220; 89 km). Near Mare Nectaris is a striking trio of craters — Theophilus (268; 100 km), Cyrillus (94; 93 km), and Catharina (76; 97 km). Theophilus is 4,400 metres deep and the central mountain chain is up to 2,000 metres high.

Located along 60° longitude East is a row of four large craters: Langrenus (174; 132 km), Vendelinus (273; 147 km), Petavius (214; 177 km) and Furnerius (126; 125 km). West of Furnerius is the crater Rheita (240; 70 km), adjacent to the well-known crater valley Vallis Rheita (u). Inside the large walled plain Janssen (163; 190 km) there is a broad rille.

On the southeastern limb of the Moon one can see part of Mare Australe (A), which extends onto the far side of the Moon. Mare Smythii (Sm), on the eastern limb of the Moon, appears markedly distorted because of its position near the limb; in reality it is circular.

age of Moon 1 d—7 d,
15 d—21 d

name index see p. 181 ff.
— · — · — clefts

*1752 Lacaille and Lalande (France) measured the distance of the Moon from the Earth.*
*1946 First radar reflections from the Moon obtained in the USA.*
*1969—1970 First measurement of the Moon's distance from the Earth by means of lasers (precision: metres, later tens of centimetres).*

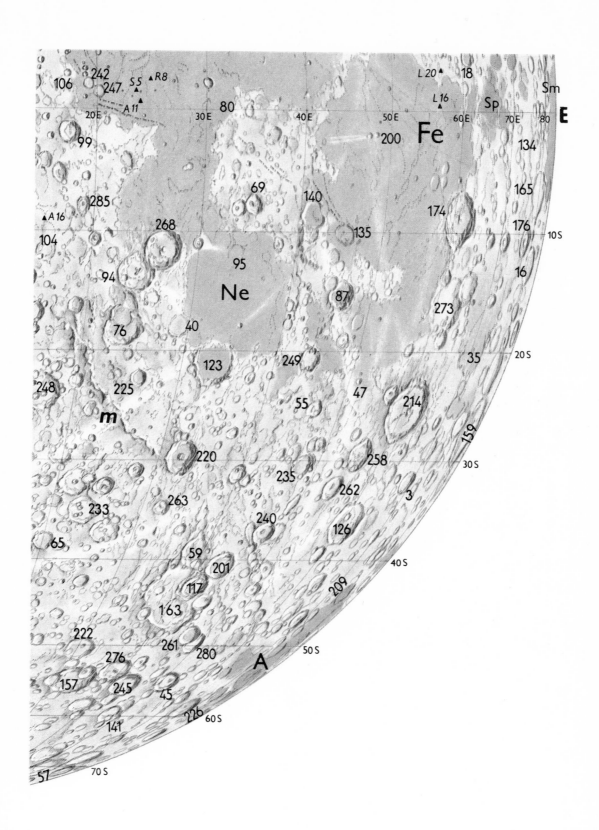

242
106
247
S5
R8
A11
20E
80
30E
40E
L20
18
Sm
L16
Sp
50E
60E
70E
80
E
99
200
Fe
134
69
140
165
285
174
176
10S
A16
135
104
95
16
94
Ne
87
273
76
40
35
20S
123
249
159
248
225
55
47
214
m
258
30S
220
235
262
3
233
263
240
126
65
59
201
209
40S
117
163
222
261
280
A
50S
276
245
45
157
226
60S
141
57
70S

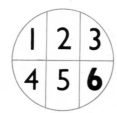

| 1 | 2 | 3 |
|---|---|---|
| 4 | 5 | **6** |

**L6**

The Sun is a star with a diameter of 1,392,000 kilometres (109 times the diameter of the Earth). In its interior, its core (a), at a temperature of approximately 14 million degrees, a nuclear reaction whereby hydrogen is converted into helium is constantly taking place and producing energy. The energy released in the core flows most of the way outward to the surface in the form of radiation (b) and then in the outer layers of the Sun is distributed to the surface by the movement of gases, by convection (c).

The visible surface of the Sun is the surface of the *photosphere* (d), which is approximately 300 kilometres high, has a temperature of 6,000° and emits intense radiation in the visible region of the spectrum. In the photosphere we often see dark sunspots (e) and bright faculae (f). For more about the photosphere see p. 60.

The region above the photosphere, extending to a height of 10,000 kilometres, is called the *chromosphere* (g). Spurting upward from its surface are numerous finger-like jets — spicules (h). High above the chromosphere rise luminous clouds of gas — *prominences*. These may be observed as protuberances on the edge of the solar disc (k) or as dark *filaments* (m) on the solar disc. Active regions are the site of occasional sudden flares — *chromospheric eruptions* (n), which are a source of high energy radiation that manifests itself also on the Earth (magnetic storms, auroras, and so on).

The outermost part of the Sun's atmosphere is called the *corona* (c).

photosphere — see p. 60
chromosphere and prominences — see p. 62

corona:

during a total eclipse
   of the Sun
(see astronomical year-book)

*1814—1815 Frauenhofer (Germany) — description of the absorption lines in the solar spectrum.*

*1868 Lockyer (England) — discovery of helium on the Sun.*

*1891 Hale (USA) and Deslandres (France) introduced the spectroheliograph — observation of the chromosphere.*

*1905 Wien (Germany) determined the temperature of the Sun's surface (about 6,000° C).*

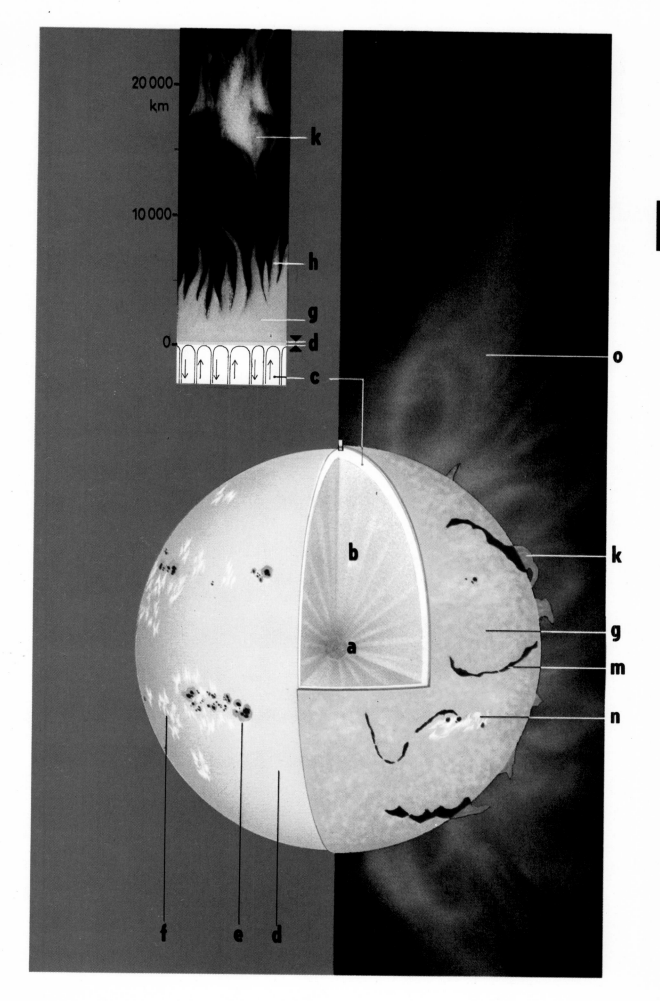

20 000 km

10 000

0

k

h

g

d

c

b

a

f

e

d

o

k

g

m

n

# THE PHOTOSPHERE
## SUNSPOTS

When we look at the Sun, it is very important to provide effective protection for the eyes and also prevent heating of the instrument optics. Ordinary smoked glass is not enough. The Sun can be safely observed even without special filters and the like by *projection* through the eyepiece of a telescope (or binoculars) as shown in Figure a. Project the Sun's image on a sheet of white paper and focus it by moving the eyepiece. What we 'see' when we observe the Sun is radiation from the photosphere.

The brightness of the solar disc diminishes towards the perimeter. *Sunspots* are areas with a temperature as much as 2,000° less than that of the surrounding photosphere. They occur in so-called active regions, where they form *spot groups* (b) up to 100,000 kilometres in diameter. A developed sunspot is composed of a dark umbra and lighter penumbra with filamentous structure. The development of large spot groups and their subsequent fading occurs in the sequence A to J as shown. Sunspots provide landmarks on the Sun and make it possible to observe the rotation of the Sun round its own axis (once in 27.3 days). Peak levels of sunspot activity are reached every 11 years or so, with quiet years midway between the peaks.

Clouds of hot gases rising from the convective layer (p. 58) produce a granular structure in the visible photosphere — *granulation.* The individual granulae measure approximately 1,000 to 2,000 kilometres in diameter.

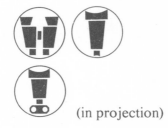

(in projection)

daily
WARNING!
  Protect the eyes!!!

a — projection of the Sun's image through an eyepiece
b — spot group
c — detail of a sunspot and granulation (the blue ring denotes the Earth)
A—J Zurich classification of spot groups

*1611 Fabricius (Holland) — discovery of the rotation of the Sun according to the apparent movement of sunspots on the solar disc.*
*1844—1851 Schwabe (Germany) — discovery of periodicity in the occurrence of sunspots.*

A

B

C

D

E

F

G

H

J

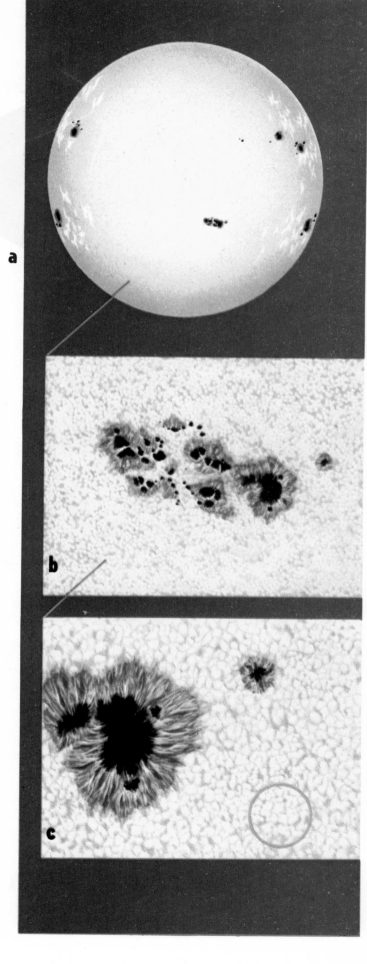

a

b

c

Prominences may be compared to the clouds in the Earth's atmosphere; on the Sun these are relatively cool hydrogen clouds in the corona with a density a hundred to a thousand times greater than that of the corona. The occurrence of prominences and their development is closely connected with the interactions between the electrically charged plasma and magnetic fields on the Sun.

A typical prominence is an elongate, flat formation 100,000 to 200,000 kilometres long, 20,000 to 40,000 kilometres high and 5,000 to 10,000 kilometres thick. Only part of this formation is visible at the edge of the Sun (a, b). On the solar disc prominences appear as dark filaments (b, c). A *quiescent prominence* (a) remains in the same spot without any great changes for weeks and even months. Prominences of the *surge* type (d), expelled from the chromosphere at a speed of several tens of kilometres per second, are marked by very rapid development. In *loop prominences* (e) gas moves along the lines of magnetic force.

Prominences radiate only in certain wavelengths (spectral lines), notably in the red light of hydrogen H-alpha ($\lambda = 656$ nm). For observation at times other than during a solar eclipse it is therefore necessary to use a special instrument (a prominence spectroscope or coronograph).

during a total eclipse — see astronomical year-book
daily: with a prominence telescope

a — quiescent prominence
b, c — limb prominence and filament
d — surge prominence
e — loop prominence

1842 *Prominences observed for the first time.*
1858 *Prominences photographed for the first time.*
1868 *Lockyer (England) and Janssen (France) — discovery of the spectroscopic method of observing prominences at times other than during an eclipse.*

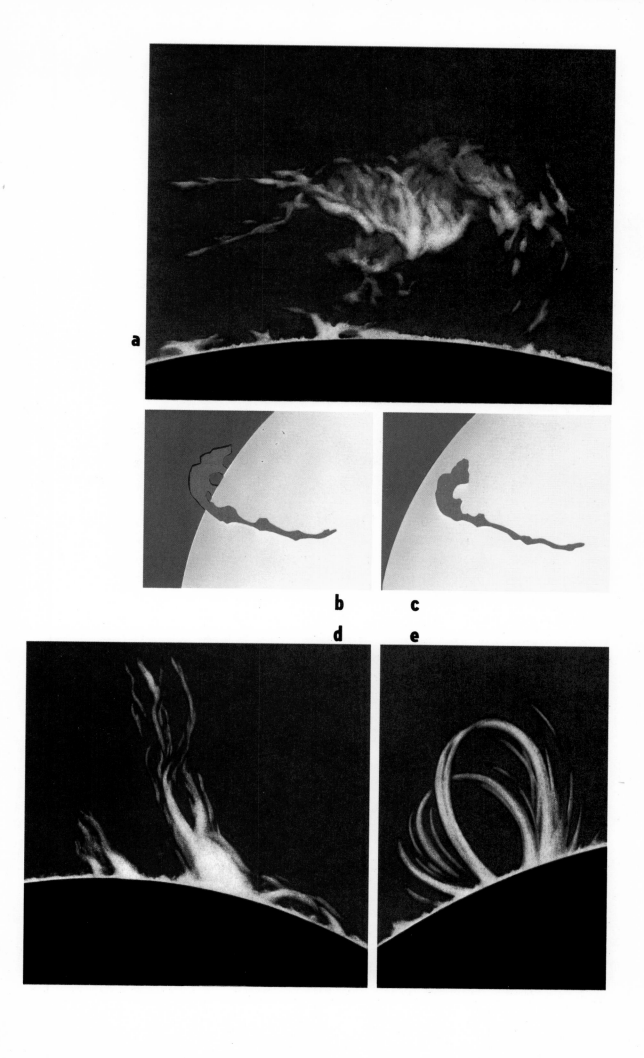

# THE SOLAR SYSTEM

The Sun is the central body of a system containing 9 known planets and their satellites (more than 30 in all), several tens of thousands of minor planets (asteroids), comets, meteor streams and individual meteoritic bodies, interplanetary dust, gas and various particles (ions, free electrons, and so on).

The plate on the opposite page gives an idea of the dimensions of the planetary system and its inhabitants. Planets are divided according to their relationship to the Earth's orbit about the Sun into two groups — the *inferior planets* (Mercury and Venus), which lie closer to the Sun than Earth, and the *superior planets* (from Mars to Pluto), outside Earth's orbit. In the plate the distances are given in millions of kilometres ($10^6$ km) and *astronomical units* (A.U.) — that is, in multiples of the distance Earth—Sun. The orbits of planets are almost circular ellipses; greater eccentricity (deviation from circular orbit) is exhibited by the orbits of Mercury, Mars and above all Pluto, shown in detail at bottom right together with the orbit of Neptune. The period of revolution round the Sun in years is given for all orbits.

Shown at the left are the relative dimensions of the planets (b) and their satellites (a) compared with a section of the solar disc. The great differences in diameter between the terrestrial planets (c) and the major planets (d) are clear; they also differ in chemical composition (see p. 14).

Observation:
on the planets see p. 70—99

signs of the planets — see
  p. 19
a — planet satellites
b — planets and their diameters in km
c — terrestrial planets
d — major planets (apart from Pluto)

1687 Newton (England) — formulation of the law of gravity.
1755 Kant and 1796 Laplace (France) — fundamental hypotheses on the origin of the solar system.
1799 to 1825 Laplace (France) — the five-volume Mécanique céleste (Celestial Mechanics).

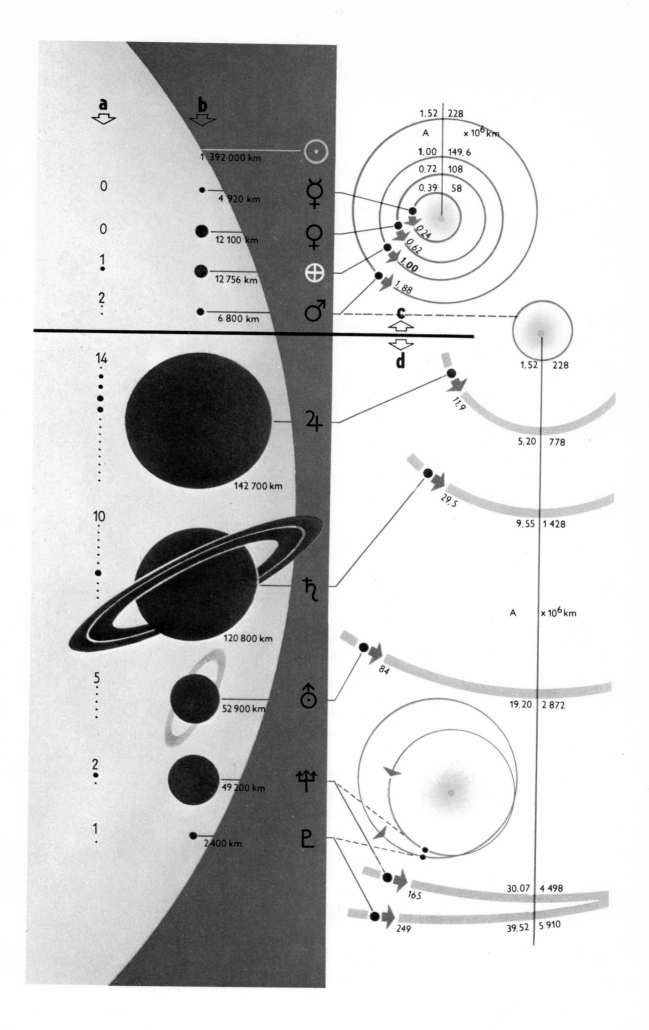

**a**   **b**

1 392 000 km   ☉

0        4 920 km   ☿

0        12 100 km   ♀

1        12 756 km   ⊕

2        6 800 km   ♂

**c**

**d**

14       142 700 km   ♃

10       120 800 km   ♄

5        52 900 km   ♅

2        49 200 km   ♆

1        2 400 km   ♇

1.52   228
A        × 10⁶ km
1.00   149.6
0.72   108
0.39   58

0.24
0.62
1.00
1.88

1.52   228

11.9
5.20   778

29.5
9.55   1 428

A        × 10⁶ km

84
19.20   2 872

165
30.07   4 498

249
39.52   5 910

# APPARENT MOTION
## OF THE SUN
## AND PLANETS

The movement of the Earth round the Sun may be observed through the apparent motion of nearby celestial objects.

Simplest is the annual movement of the Sun along the ecliptic e (top picture). The ecliptic is the apparent annual path of the Sun on the celestial sphere. In one day the Sun moves approximately 1° eastward in the opposite direction to the apparent rotation of the sky. In one year the Sun passes along the ecliptic through the 12 zodiacal constellations (maps on pp. 145—155). That is why the visibility of the constellations changes during the four seasons of the year.

The apparent motion of the planets is far more complex. The bottom picture shows how a loop is formed in the apparent orbit of a superior planet (such as Mars). The corresponding positions of the Earth and planet in their orbits and the position of the planet in the sky are marked with the same number. The planet generally moves past the background stars in an eastward direction *(direct motion)*. But in the phase when the more rapidly moving Earth begins to 'pass' the planet, the planet halts its eastward motion (at the stationary point — 5) and then changes direction in *retrograde motion.* This is followed by a second stationary point (7) and return to direct motion.

Loops in planetary orbits may be tracked on a star map on the basis of observation or data from an astronomical year-book.

tracking apparent motions of planets;
see also astronomical year-book and the diagrams on pp. 161—175.

top — apparent motion of the Sun along the ecliptic (e)
bottom — apparent motion of a superior planet

*2nd millenium B.C. Babylon — introduction of the circle of zodiacal constellations.*
*about 140 A. D. Ptolemaios (Alexandria) — formulated the theory of the geocentric system. Movement of the planets in epicycles.*

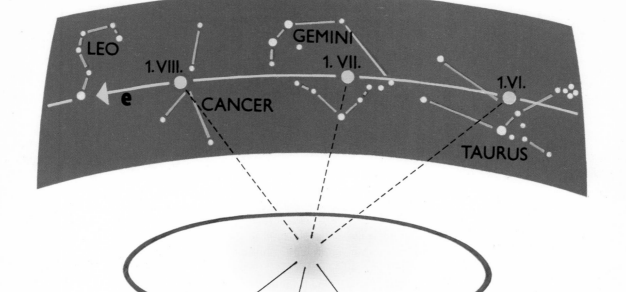

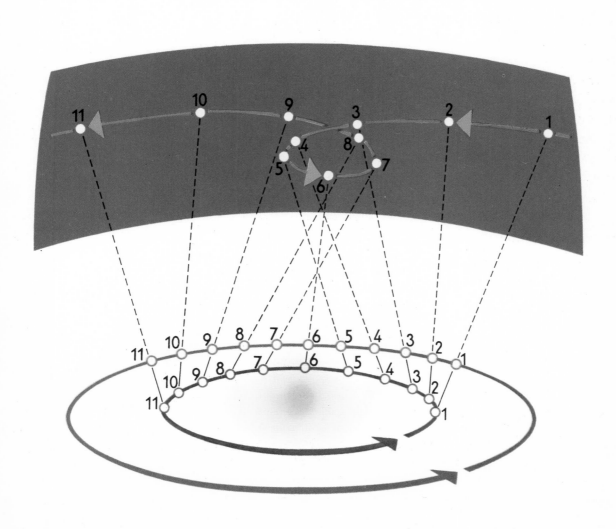

# CONJUNCTION, OPPOSITION, ELONGATION, QUADRATURE CONFIGURATION OF THE PLANETS

Since the distant past *astrology* has claimed to foretell a person's future according to the positions of the planets, and to this day we use the terms and symbols of astrology to designate the various *aspects*, the important positions of the planet in relation to the Sun. Knowledge of the aspects is useful for quick orientation in observation, but this has nothing at all to do with astrological belief in the significance of 'heavenly signs'.

The main aspects of an inferior planet (Figure a) are: inferior conjunction ($C_2$), superior conjunction ($C_1$), eastern elongation (EE) and western elongation (EW). (See also p. 72).

A superior planet (Figure c) cannot be observed when it is in conjunction with the Sun (C) and is visible all night when it is at opposition (O). At quadratures (Q) the angle between the planet and the Sun, as seen from the Earth, is 90°.

Striking and easily observed astronomical phenomena occur at special configurations of the Moon and planets, particularly their conjunction. Figure b and the adjacent drawing at the right depict the conjunction of the Moon M and planets $P_1$ and $P_2$ in the morning sky. Those objects which appear so close together on the sky, however, are in fact hundreds of millions of kilometres apart, along the 'line of sight'.

Conditions for observing the five brightest planets and data pertaining to their aspects in the years 1979 to 2000 are shown in the tables on pp. 161—175.

tracking the configurations of the planets and the Moon — see also astronomical year-book and the diagrams on pp. 161—175

a — aspects of an inferior planet
b — conjunction of two planets and the Moon
c — aspects of a superior planet

1543 *Copernicus (Poland) — hypothesis that the Earth and other planets revolve around the Sun (heliocentric system) expounded in the book* De revolutionibus... (On the Revolutions...).

1728 *Bradley (England) — discovery of stellar aberration — proof of the Earth's orbital motion round the Sun.*

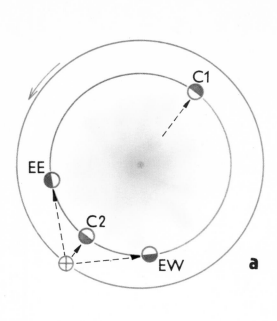

a

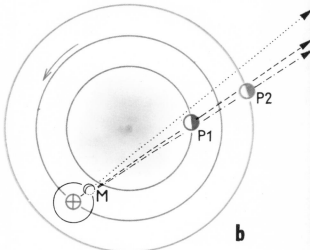

b

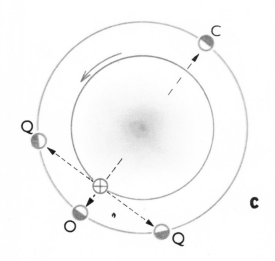

c

# PLANETS
## THROUGH THE TELESCOPE
## DIAMETER, COLOUR,
## BRIGHTNESS

Telescopic observation of the planets places great demands on the quality of the telescope and experience of the observer. The angular diameters of the planets are at most several tens of seconds of arc. On the opposite plate the planets are drawn in the same scale as they appear when viewed through a telescope.

At the top we see the size of the discs of Mercury and Venus in their various phases (p. 73). At right, in the same scale, is a lunar crater 50 kilometres in diameter (the whole Moon would have a diameter of approximately 1,5 metre on this scale). Next comes Mars close to conjunction with the Sun (smallest disc) at aphelion (centre) and at perihelion (at left). (See also p. 79). The discs of Jupiter and Saturn correspond to the greatest and least distance of these planets from the Earth.

At the bottom is a graphic depiction of the corresponding changes in apparent stellar magnitude of the planets (in the case of Mars from $+1.6^m$ to $-2.8^m$) and their colour in the sky.

If this plate is placed at a distance of about 165 metres, the discs of the planets will be at the same angles as in the sky (suitable for testing both telescope and observer!).

observe the plate through a telescope at a distance of 165 metres and draw what you see on the disc of Mars, Jupiter, etc.

symbols of the planets — see p. 19.

*till the end of the 19th century. Observation of the surface of planets, drawing of details.*

*1st half of the 20th century. New methods of observation, beginnings of planetary physics.*

*1962 Commencement of planetary research with space probes.*

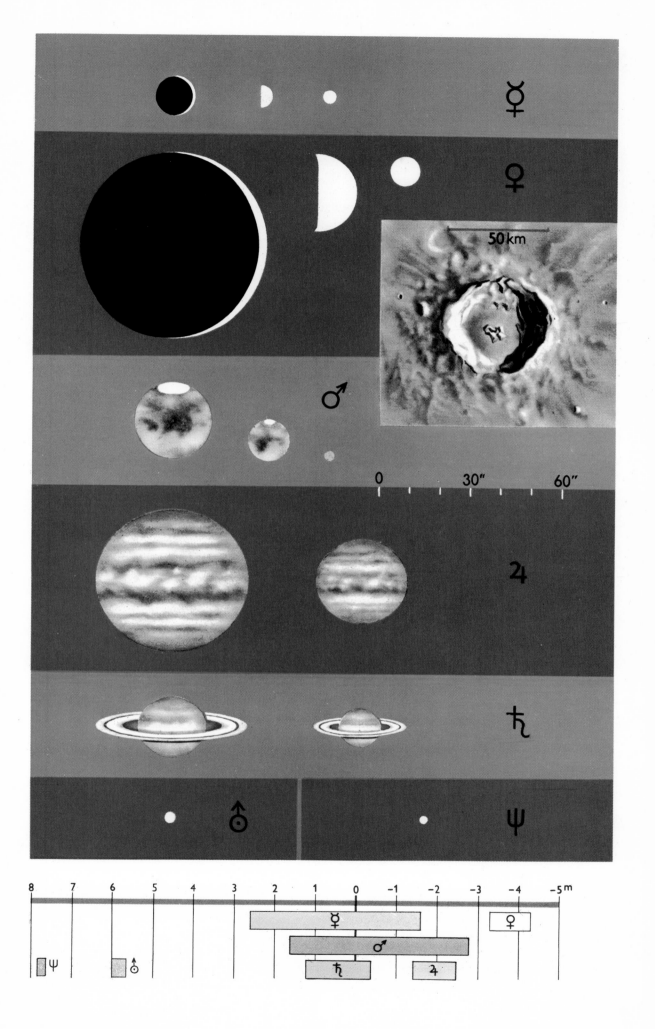

The ancient Greeks believed that the eastern and western elongation of Mercury were marked by two different objects which they named Hermes (the evening star) and Apollo (the morning star), respectively. When they later discovered that the two were one and the same planet they retained the first of the two names — Hermes (called Mercury by the Romans).

Mercury is an inferior planet that never elongates more than 28° from the Sun, which limits its visibility from the Earth to about one hour before sunrise or after sunset and then only when it is at its greatest elongation.

When viewed from the Earth (a), the side of Mercury illuminated by the Sun may be observed in various phases, as with the Moon and Venus. In position 1 Mercury is at inferior conjunction and cannot be observed. Later it appears to the right of the Sun in the morning sky as a *morning star* (b). In position 2 it is at greatest western elongation, then it again approaches the Sun, passes through superior conjunction (3) and moves east of the Sun to the evening sky, where it appears as an *evening star* (c). Mercury is most likely to be observed at the time of its greatest eastern elongation (4) when it is 18° to 28° from the Sun.

The planet's disc has a faint pinkish tinge (d). Besides phases it is possible, on rare occasions, to observe further details — greyish spots — with a larger telescope. The renowned French astronomer and observer of planets A. Dollfus, stated that under excellent conditions of observation Mercury (viewed through a large telescope) resembles the Moon observed with the unaided eye. Dollfus's observations were also used in the construction of the albedo map on the following page.

A solar day on Mercury is equivalent to 176 terrestrial days. Mercury has no atmosphere or water. Life is not believed to exist there, not even in the form of microorganisms.

only during periods of greatest elongation — see astronomical year-book and the diagrams on pp. 161—175

a — origin of phases of an inferior planet
b — Mercury as a morning star
c — Mercury as an evening star
d — Mercury as it looks through a telescope

*1639 Zupus (Italy) — first observations of the phases of Mercury.*
*1800 Schröter and Harding (Germany) — first records of the observation of dark spots on Mercury.*

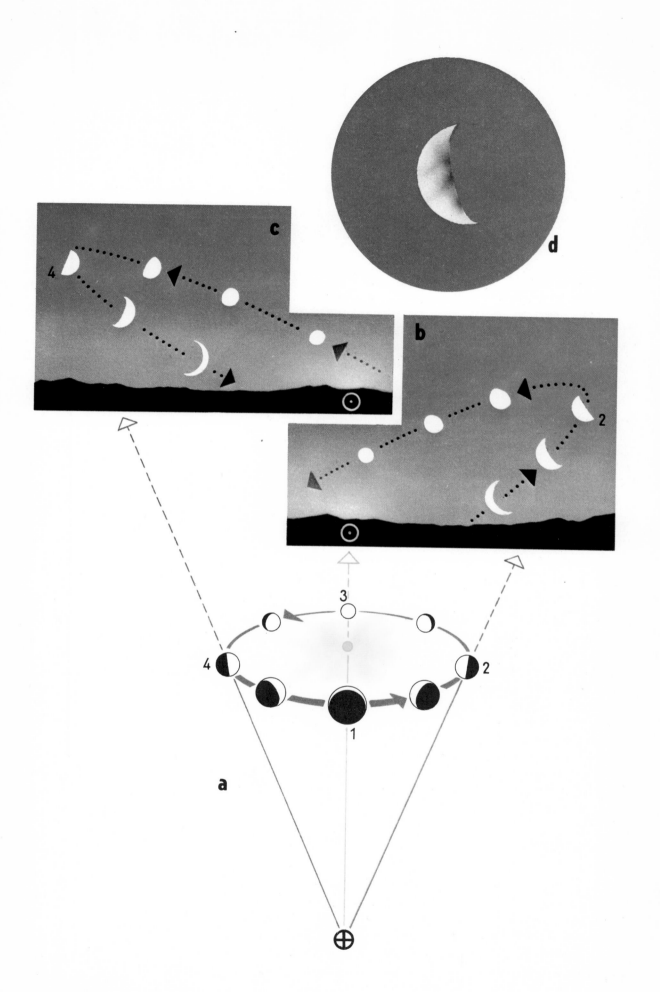

a

b

c

d

1

2

3

4

# MAPS OF MERCURY ALBEDO AND TOPOGRAPHIC MAPS

The light and dark areas on Mercury are far less pronounced than the main albedo formations on the Moon or Mars, and it is extremely difficult to map them from Earth. The names on the albedo maps shown here are those approved by the International Astronomical Union.

Detailed investigation of the surface of Mercury was begun in the years 1974—1975, when the US space probe Mariner 10 flew past the planet three times and transmitted to Earth television photos and the results of physical measurements. The best photographs made by Mariner 10 have a resolving power of 50—100 metres, which is approximately 10,000 times greater than can be attained from the Earth. The space probe passed over the day as well as the night side of Mercury, always under identical conditions of illumination by the Sun; in all, it succeeded in mapping in detail approximately 40 per cent of the planet's surface.

It appears that the classic dark regions on Mercury are in substance vast plains irregularly pockmarked with craters. Light areas appear in the region of bright rays or groups of bright-rayed craters, which, of course, cannot be resolved with terrestrial telescopes. In this sense the spots on Mercury are light spots on a dark ground rather than dark spots on a light ground (such as the 'mare' on the Moon). 'Seas' on Mercury are called 'planitiae' (plains).

The surface of Mercury is a pendant of the lunar surface in another part of the Solar System and on a planet that has a far greater mass than the Moon. We find here numerous craters, faults, low and rounded ridges, relatively smooth intercrater plains between large craters, basins, rayed craters, and so on. To date, the largest known plain on Mercury is Caloris Planitia (Plain of Heat), measuring approximately 1,300 kilometres in diameter.

For nomenclature on the maps of Mercury see p. 20.

*1965 Pettengill, Dyce, Colombo (USA) — period of rotation of Mercury (58.6 days) measured with radar.*

*1974—1975 Mariner 10 spacecraft (USA) flew past Mercury three times. Mapping of the surface, discovery of a magnetic field and so on.*

albedo formations may be observed only on rare occasions with a large telescope

On the albedo maps the hemisphere of Mercury illuminated by the Sun and mapped by Mariner 10 is marked in yellow.

S. — Solitudo, PL. — Planitia

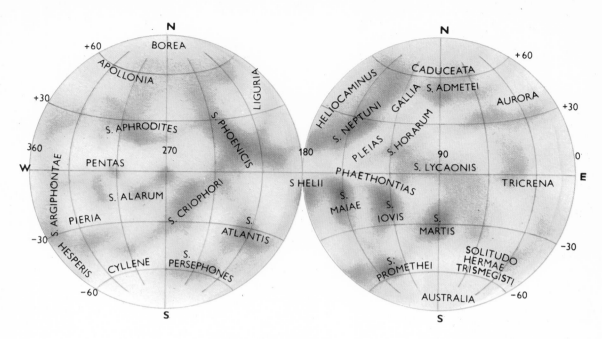

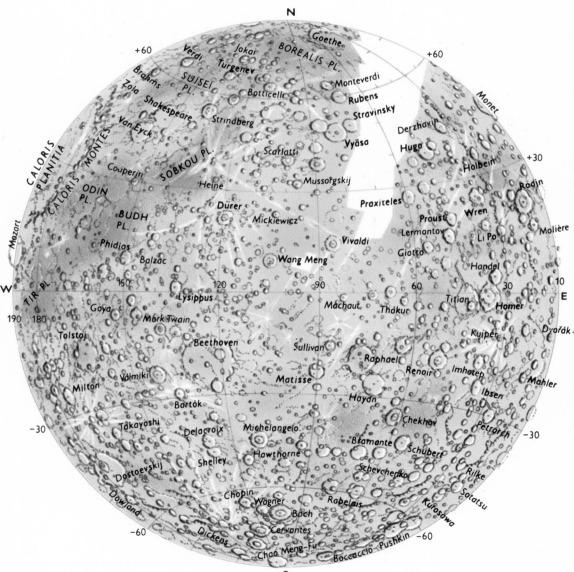

# VENUS
## VIEWED
## FROM THE EARTH
## AND FROM
## A SPACESHIP

Venus is the third brightest celestial body after the Sun and Moon; under favourable conditions it can be seen with the unaided eye even in the clear sky during the day. Like Mercury it, too, appears as a morning or evening star (p. 73) but it elongates up to 47° from the Sun, so that it shines late into the night.

The phases of Venus may be observed with a small telescope. But even with larger instruments little more is revealed than information about a phase (b). The ill-defined, greyish spots and bands seen by some observers (a) are more likely the effect of the marked contrast between the brightly lit part of the planet and the background.

From the Earth one can observe only the top of the impenetrable layer of clouds about 65 kilometres above the surface of Venus. Above this, to a height of 76 kilometres, is another transparent layer which in ultraviolet (UV) radiation appears brighter than the dense cloud layer beneath. The structure of the clouds was already apparent on UV photographs made from the Earth (c) but it was the UV photographs sent back by Mariner 10 in 1974 that first made it possible to distinguish details. The cloud layer rotates round the planet's axis once in 4 days but the period of rotation of the planet itself is 243 days.

The atmosphere on Venus is composed primarily of carbon dioxide. The pressure on the planet's surface is 90 times greater than the atmospheric pressure on the surface of the Earth and the surface temperature is as much as 470° C.

(phases)

daily except for periods of conjunction; see astronomical year-book and the diagrams on pp. 161—175

appearance of Venus:
a, b — through a telescope
c — on a UV photograph
d — on a photograph made by Mariner 10

1610 Galileo (Italy) — discovery of the phases of Venus.
1961—1966 Radar measurements that led to the determination of the period of rotation of Venus (243 days).
1970 Venus 7 space probe (USSR) — first soft landing on Venus, recorded the temperature (475° C) and pressure (9 MN/m²) on the planet's surface.

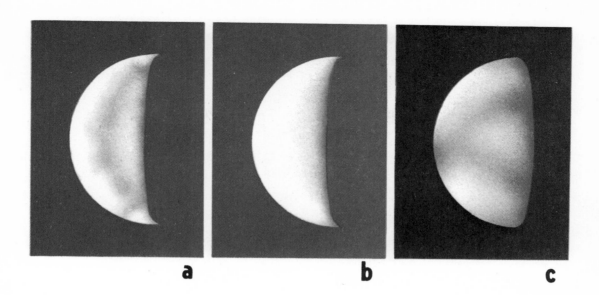

a　　　　　　　b　　　　　　　c

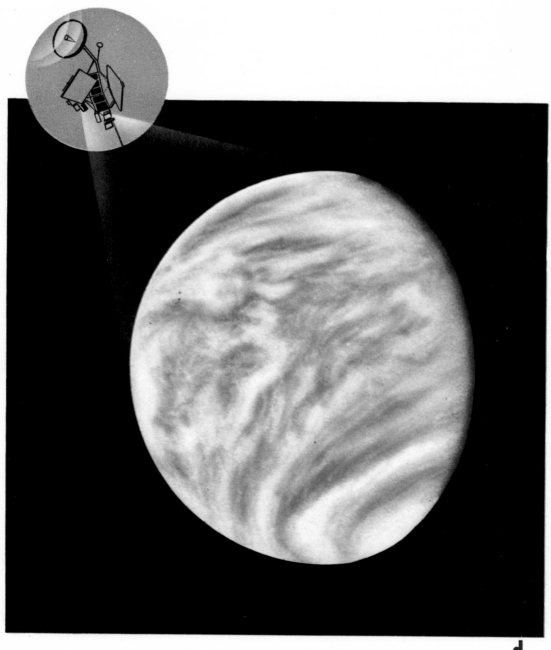

d

# MARS
## ANGULAR DIAMETER
## AT OPPOSITION

The best time for observing Mars with a telescope is during the several-month period round opposition when the Earth is between Mars and the Sun and Mars is closest to Earth. Mars is at opposition about every 780 days, as shown by the series of five diagrams at the top (Earth is coloured blue, Mars red).

Mars' orbit round the Sun is markedly eccentric. At perihelion (P) Mars is 207 million kilometres distant from the Sun and at aphelion (A) it is 249 million kilometres from the Sun. The distance between Mars and the Earth (and also Mars' angular diameter in the sky) at opposition depends, therefore, at what point in Mars' orbit opposition occurs. When Mars is in opposition at perihelion, it comes to within 56 million kilometres of the Earth and its diameter may be as much as 25″. Oppositions close to these optimum conditions will occur in 1986 and 1988.

Mars' axis of rotation is tilted at a constant angle of 25° to its orbital plane. This produces changes in seasons similar to those on Earth, as Mars alternately presents its northern and southern hemispheres towards the Sun, depending on its position in orbit. All this affects the appearance of the planet through the telescope.

oppositions of Mars with the Sun

blue — orbit of the Earth

red — orbit of Mars

P — perihelion

A — aphelion

Roman numerals — date on Earth

the number beside each opposition gives the distance between the Earth and Mars in millions of kilometres.

1609 *Kepler (Germany) — discovery of the first two laws of the motion of planets based on the analysis of observations of the positions of Mars at various oppositions (observations by Tycho Brahe).*

1666 *D. Cassini (Italian in France) — discovery of the rotation of Mars.*

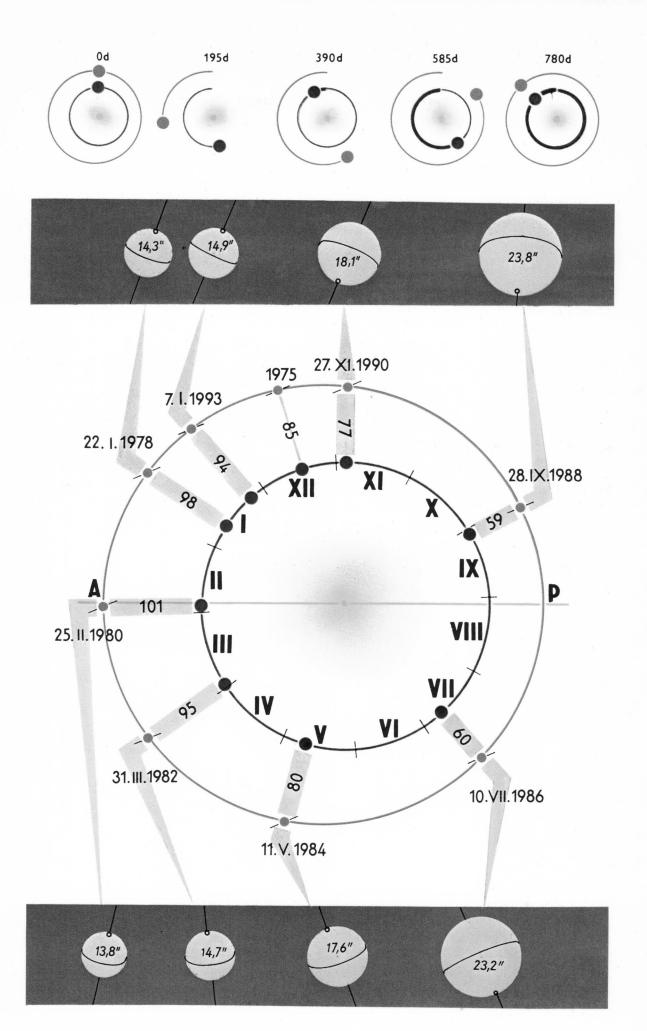

The series of maps M 1—M 4 shows Mars as it appears to the terrestrial observer through a telescope. North is at the top, as is the rule on maps of planets; with a reversing telescope the image will be upside down. The four maps show four different regions of Mars, with the planet rotated by 90° in each consecutive map.

Just as geographical coordinates serve to define positions on the Earth, so coordinates are used to define positions on Mars — namely *areographic longitude* L and *areographic latitude* B. For the observer it is important to know the areographic longitude of the central meridian, the meridian which at the moment of observation passes through the centre of the planetary disc. This is given in astronomical year-books. To identify the various details it is also necessary to take into account the season on Mars and the corresponding orientation of its rotational axis. At the same length L of the central meridian known details may appear quite different, depending on whether the North Pole N (map bottom left) or South Pole S (map bottom right) is turned towards the Earth. For purposes of better orientation, therefore, small maps corresponding to the extreme position of the North and South Poles in relation to Earth have been added to Maps M 1—M 4.

during periods of opposition — see p. 79

Details on the map:
1 — Oxia Palus
2 — Hyperboreus Lacus
3 — Lunae Lacus
4 — Juventae Fons
5 — Coprates
6 — Melas Lacus
see also maps M 4 and M 2

*1783 W. Herschel (England) stated that of all the planets Mars most greatly resembles the Earth (he observed the melting of the polar caps).*
*1830—1839 Beer and Mädler (Germany) — first map of Mars.*

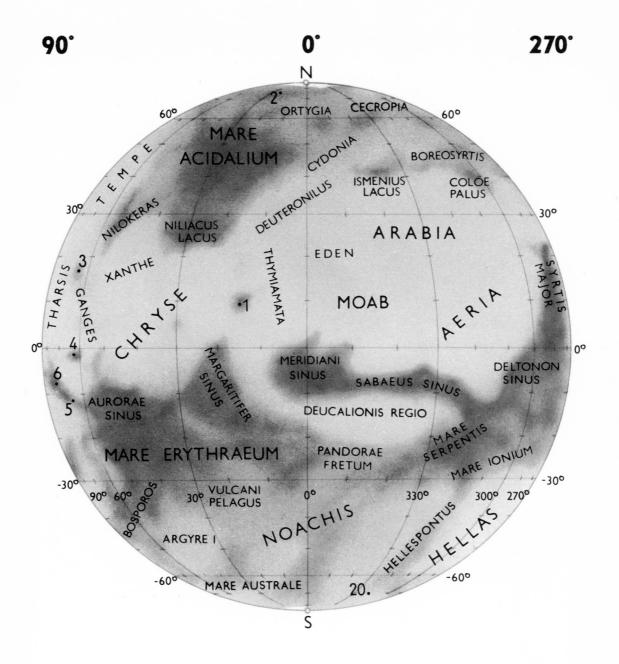

N

60°   2°   ORTYGIA   CECROPIA   60°

MARE
ACIDALIUM    CYDONIA    BOREOSYRTIS

30°   ISMENIUS   COLÖE
LACUS   PALUS

T E M P E

NILOKERAS   DEUTERONILUS   ARABIA   30°

NILIACUS
LACUS   EDEN

THARSIS   .3   XANTHE   THYMIAMATA   MOAB   AERIA   SYRTIS MAJOR

GANGES   CHRYSE   .1

0°   .4   0°

.6   MARGARITIFER SINUS   MERIDIANI SINUS   DELTONON SINUS

.5   AURORAE SINUS   SABAEUS SINUS

DEUCALIONIS REGIO

MARE ERYTHRAEUM   PANDORAE FRETUM   MARE SERPENTIS   MARE IONIUM

-30°   BOSPOROS   VULCANI PELAGUS   -30°

90° 60°   30°   0°   330°   300° 270°

ARGYRE I   NOACHIS   HELLESPONTUS   HELLAS

-60°   MARE AUSTRALE   20.   -60°

S

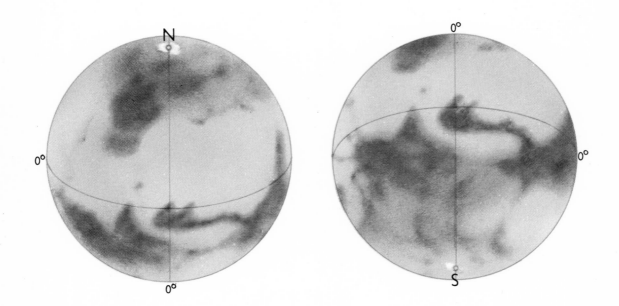

Observation of the details on the surface of Mars is among the most difficult of all astronomical observations and a beginner lacking the necessary experience will see only the most striking formations.

The surface formations which can be observed through the telescope are divided into three basic types, differing by their albedo and colouring. *Albedo* characterizes the reflecting power of various surfaces and is expressed as a ratio of reflected light to the total amount falling on the surface. The phenomena on Mars with the highest albedo are the dazzling white polar caps. Slightly darker are the typical orange areas of the 'continents' and the areas with the lowest albedo are the dark 'seas' (mare), 'lakes' (lacus), 'bays' (sinus), and the like.

The *polar caps* can be seen well even with a small telescope, particularly during the winter season on the Martian hemisphere facing the Earth, when the cap grows to 60° latitude or even farther towards the equator. In summer all that remains of the cap is a small remnant at the pole; this 'summer cap' (composed of frozen water) is shown as a white patch on the maps.

during periods of opposition
— see p. 79

Details on the map:
 7 — Nectar
 8 — Meotis Palus
 9 — Phoenicis Lacus
10 — Sirenum Sinus
11 — Gorgonum Sinus
see also maps M 1 and M 3

*1877 Schiaparelli (Italy) — discovery of canals on Mars; later hypothesis as to their being of artificial origin.*
*1909 Antoniadi (France) — observational proof refuting the geometric character of the canals.*

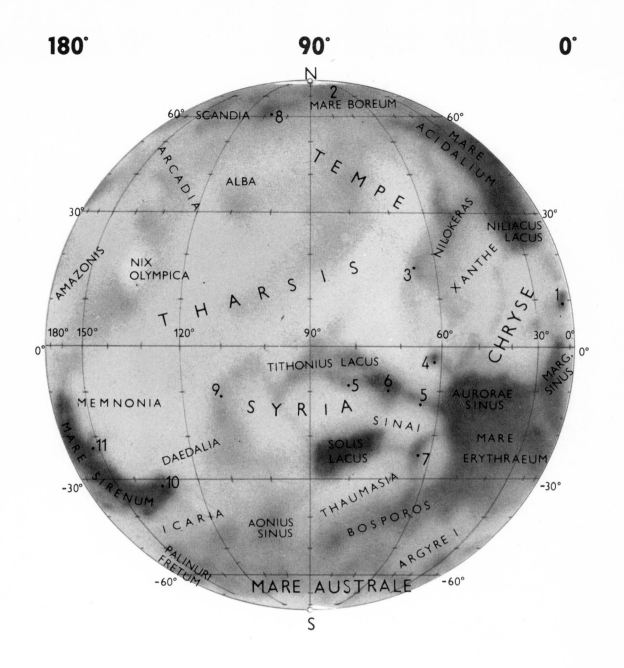

180°   90°   0°

N
2
MARE BOREUM
60°   SCANDIA   •8   60°
MARE
ACIDALIUM
ARCADIA
30°   NILOKERAS   NILIACUS   30°
ALBA   LACUS
TEMPE
AMAZONIS   NIX   XANTHE
OLYMPICA   3•
THARSIS
180° 150°   120°   90°   60°   30°   0°
0°   CHRYSE   1•   0°
TITHONIUS LACUS   4•   MARG.
9•   5•   6•   SINUS
MEMNONIA   SYRIA   5•   AURORAE
SINAI   SINUS
•11   SOLIS   MARE
DAEDALIA   LACUS   7•   ERYTHRAEUM
MARE   •10
SIRENUM   THAUMASIA
-30°   ICARIA   BOSPOROS   -30°
AONIUS   ARGYRE I
SINUS
PALINURI
FRETUM   -60°   -60°
MARE AUSTRALE
S

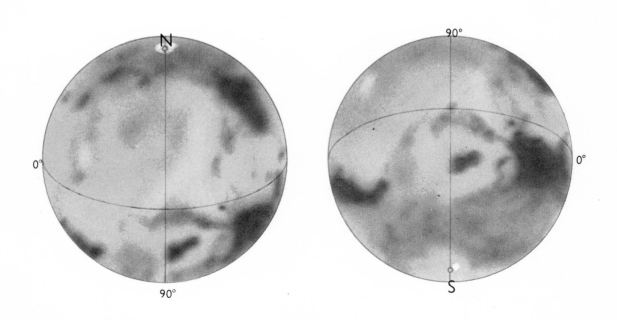

N   90°
0°   0°
90°   S

Dark albedo formations cover about one-third of Mars' surface. The meaning of their Latin names are as follows:

mare — sea, lacus — lake, palus — marsh, sinus — bay,

fretum — strait, promontorium — cape, fons — spring,

regio — landscape, depressio — depression, insula — island.

Unlike the dark 'seas' on the Moon, the dark formations on Mars undergo changes in shape, size and shading caused by the shifting of dust by the wind. For this reason albedo maps show only the more stable formations. The darkest and best visible are Syrtis Major (M 1, M 4), Mare Sirenum (M 2, M 3), Sinus Sabaeus (M 1, M 4), Mare Tyrrhenum (M 3, M 4) and Mare Acidalium (M 1, M 2). Most conspicuous of the light areas is Hellas (M 1, M 4), which is sometimes mistaken for a polar cap. The light spot Nix Olympica (M 2, M 3) is in reality a white cloud which sometimes forms above the mountain Olympus Mons (p. 89).

The thin Martian atmosphere is composed primarily of carbon dioxide and the atmospheric pressure on the surface is 300 to 800 N/m², which corresponds to the density of the Earth's atmosphere at an altitude of approximately 30 kilometres. Albedo formations, however, may be temporarily obscured or changed by the frequent ground mists and clouds of water and of frozen carbon dioxide which occur even in such a thin atmosphere.

during periods of opposition
— see p. 79

Details on map:
12 — Gigantum Sinus
13 — Titanum Sinus
14 — Trivium Charontis
15 — Sithonius Lacus
16 — Nodus Laocoontis

see also maps M 2 and M 4

*1909—1910 Tichov (Russia) — use of colour filters for studying the surface of Mars. Existence of vegetation on Mars assumed possible.*
*1922 Pettit and Nicholson (USA) — first measurement of temperature at various points on the surface of Mars.*

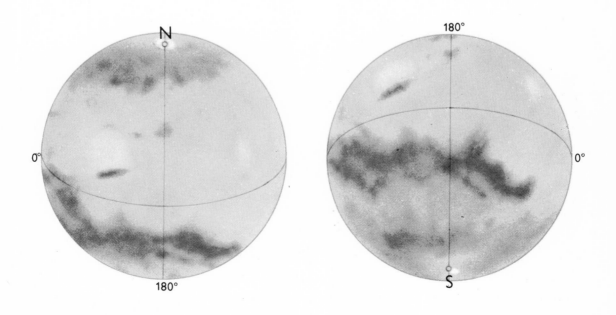

N

LEMURIA

SCANDIA

60°    15    8    60°

AETHERIA

CEBRENIA    DIACRIA

ARCADIA

CHAOS    PROPONTIS I

30°    30°

270° 240°    210°    180°    150°    120° 90°

ELYSIUM    AMAZONIS

NIX OLYMPICA

AETHIOPIS    EREBUS

16    CERBERUS    14

GOMER SINUS

MESOGAEA

THARSIS

ZEPHYRIA

0°    0°

210°    180°    150°    120°

HESPERIA

LAESTRYGONUM SINUS

13    MEMNONIA

M. TYRRHENUM

MARE CIMMERIUM

RASENA    12    9

MARE SIRENUM    11

ATLANTIS    DAEDALIA

−30°    10    −30°

AUSONIA    SCAMANDER    ICARIA

ERIDANIA    ELECTRIS

PALINURI FRETUM

−60°    MARE CHRONIUM    −60°

S

N

0°

180°

180°

0°

S

Marked changes in the appearance of albedo formations are caused by dust storms, when the velocity of the wind is as much as 400 km/h. The dust cover may spread rapidly over the whole planet, making observation of the surface impossible for weeks on end.

Even as recently as the 1950s the hypothesis that the dark areas on Mars are covered with vegetation which has a decisive influence on the seasonal changes in albedo formations still had wide support. However, after investigation by the US and Soviet space probes in 1964—1973 and particularly by the Viking 1 and 2 probes in 1976—1978 it is now certain that there is no vegetation on Mars and little probability of there being any microorganisms.

The mystery of the famous Martian 'canals' has also been solved. Through the telescope they appear as fine, dark, straight lines on a lighter surface. They were once believed to be artificial canals but later it was revealed that most were the product of visual deception; for example, they are boundaries between light and dark areas or groups of spots which the eye sees as continuous lines. One exception is the canal Coprates (M 1, M 2), which is part of the largest canyon on Mars (Valles Marineris — see p. 89).

during periods of opposition
— see p. 79

Details on the map:
17 — Nepenthes — Thoth
18 — Nilosyrtis
19 — Copais Palus
20 — Novus Mons
see also maps M 1 and M 3

*1877 Hall (USA) — discovery of Mars' satellites Phobos and Deimos.*

*1971—1972 Mariner 9 space probe (USA) — first pictures of the surface of Phobos and Deimos (small bodies of irregular shape marked with craters).*

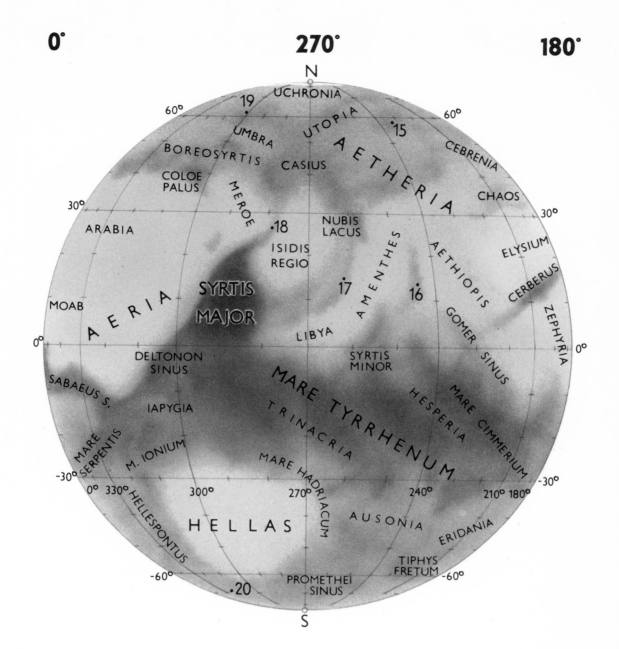

0°    270°    180°

N
UCHRONIA
19
60°                    60°
UMBRA    UTOPIA
BOREOSYRTIS    CASIUS    15    CEBRENIA
COLOE    A E T H E R I A
PALUS    CHAOS
30°    MEROE    30°
ARABIA    18
NUBIS    AETHIOPIS    ELYSIUM
LACUS    AMENTHES
ISIDIS    CERBERUS
REGIO    17    16
MOAB    SYRTIS    GOMER    ZEPHYRIA
0°    A E R I A    MAJOR    SINUS    0°
LIBYA
DELTONON    SYRTIS    MARE CIMMERIUM
SINUS    MINOR    HESPERIA
SABAEUS S.    IAPYGIA    MARE TYRRHENUM
M. IONIUM    T R I N A C R I A
-30°    MARE    MARE HADRIACUM    -30°
SERPENTIS    0° 330°    300°    270°    240°    210° 180°
HELLESPONTUS    A U S O N I A
-60°    H E L L A S    ERIDANIA    -60°
20    PROMETHEI    TIPHYS
SINUS    FRETUM
S

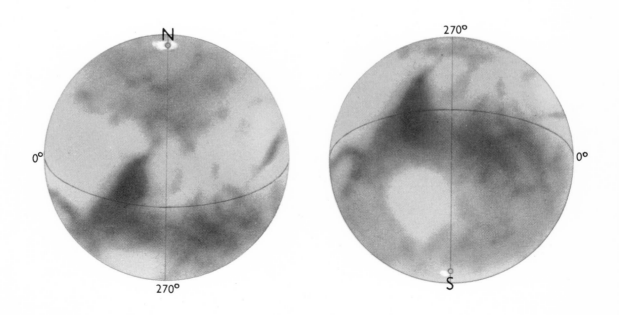

N
0°
270°

270°
0°
S

# TOPOGRAPHIC MAP OF MARS : 1
## HEMISPHERE BETWEEN AREOGRAPHIC LONGITUDES 0° AND 180°

Mapping of Mars using observations made by space probes (see opposite and next Plate) showed that albedo formations usually do not correspond to the planet's topography. Unlike the Moon, the 'seas' on Mars cover widely different types of terrain, including crater fields.

The southern hemisphere of Mars is mostly covered with craters much like the continents on the Moon. The northern hemisphere is flatter, has few craters, and is covered with basalt lava layers with numerous traces of volcanic activity. The largest volcano is Olympus Mons (detail b), a cone-shaped mountain which has a diameter of 500 kilometres at the base and is 24 kilometres high — bigger than any volcano on Earth. Impressive, also, are the dimensions of the complex of canyons Valles Marineris (detail a), which is more than 4,000 kilometres long, up to 250 kilometres wide and 6 kilometres deep.

The nomenclature for topographic maps of Mars includes besides the names of craters (after important personalities) also Latin names for various types of reliefs, e.g.:

Vastitas Borealis — a vast northern plain, Planitia — plain, Planum — plateau, Vallis — valley, Fossa — trench, Chasma — canyon, Mons — mountain, Tholus — dome-shaped hill, Patera — irregular crater, Mensa — table mountain.

One warning: Topographic regions are not identical with albedo formations of the same or similar name (compare, for example, Chryse on Map 1 and Chryse Planitia — region where Viking 1 landed — on the neighbouring map).

Relief features of Mars are visible only on photographs sent back by space probes.

Details of the surface:
a — part of the canyon Valles Marineris
b — shield volcano Mons Olympus
PL. (on the map) — Planitia

1965 *Mariner 4 space probe (USA) — discovery of craters on Mars. Measured atmospheric pressure of 400—700 N/m² on the planet's surface.*
1971—1972 *Mariner 9 space probe (USA) — 7,329 photographs made in orbit. First detailed mapping of the planet Mars.*

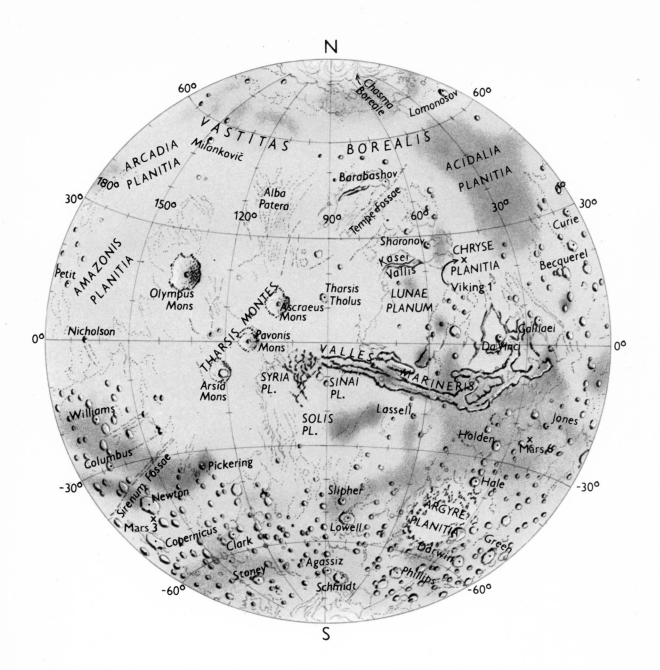

N

60°
V A S T I T A S    B O R E A L I S
Chasma Boreale
Lomonosov
60°
ARCADIA
PLANITIA
180°
150°
120°
90°
60°
Milankovič
ACIDALIA
PLANITIA
30°
30°
30°
Barabashov
Alba Patera
Tempe Fossae
Sharonov
CHRYSE
PLANITIA
Curie
Becquerel
AMAZONIS
PLANITIA
Petit
Olympus Mons
Ascraeus Mons
Tharsis Tholus
Kasei Vallis
Viking 1
LUNAE PLANUM
Galilaei
Nicholson
THARSIS MONTES
Pavonis Mons
Da Vinci
0°
0°
Arsia Mons
SYRIA PL.
SINAI PL.
V A L L E S    M A R I N E R I S
Williams
SOLIS PL.
Lassell
Jones
Columbus
Pickering
Holden
Mars 6
Sirenum Fossae
Newton
Slipher
Hale
-30°
-30°
Mars 3
ARGYRE PLANITIA
Copernicus
Clark
Lowell
Darwin
Green
Stoney
Agassiz
Phillips
Schmidt
-60°
-60°

S

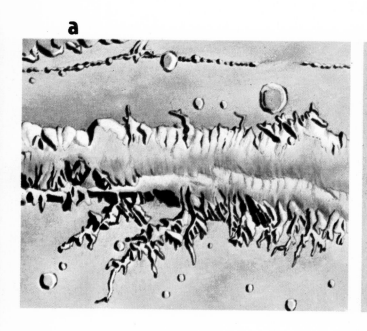

a

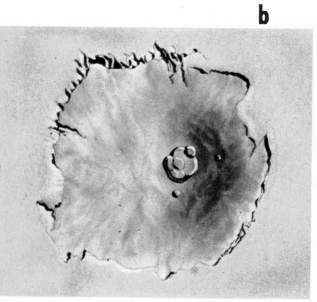

b

# TOPOGRAPHIC MAP
## OF MARS : 2
### HEMISPHERE BETWEEN
### AREOGRAPHIC
### LONGITUDES
### 180° AND 360°

It is very likely that water once coursed over the surface of Mars leaving behind numerous traces such as dry river beds (detail a) and other formations for which erosion by water appears to be the only explanation. This is surprising, for the temperature and atmospheric pressure on Mars are now so low that the presence of water on the surface can only be in the form of ice or water vapour. Conditions on the planet must have been different in the past.

At present the greatest amount of surface water is in the form of massive ice covers at the poles that do not melt even in summer. The appearance of the polar cap at the southern pole is shown in detail b. It seems that there is also a large amount of water below the surface in the form of permafrost, frozen subsoil such as is found in the polar regions on Earth.

The planet's red colour is produced by iron oxides in the surface layer. Dust particles floating in the atmosphere give the sky on Mars a pinkish tinge. The regions where Viking Lander 1 and Viking Lander 2 landed (Chryse Planitia and Utopia Planitia, respectively) were found to be desert strewn with stones and covered with dust and sand which formed low dunes in places.

Surface features of Mars are visible only on photographs sent back by space probes.

Details of the surface:
a — part of the valley Ma'adim Vallis
b — southern polar cap in summer
S — South Pole

*1974 Mars 6 space probe (USSR) — first direct measurements in Mars' atmosphere by a space probe.*
*1976 (continuing) Viking 1 and Viking 2 (USA) — complex investigation by two artificial satellites of Mars. First photographs, biochemical analyses, etc., by two modules that made a soft landing on Mars.*

N

60°                                                              60°
Koralev

V A S T I T A S    B O R E A L I S
Lyot                                                    Phlegra Montes
Deuteronilus                          Viking 2×  Mie
Mensae          Moreux          UTOPIA  PLANITIA    Hecates
30°                                                  Tholus        30°
Flammarion                                          PLANITIA
Pasteur   Cassini    Antoniadi              ELYSIUM    Elysium
Henry                          SYRTIS    ISIDIS        Mons
Araga          MAJOR    PLANITIA   Hephaestus  Albor        Orcus
0° 0°                          PLANITIA          Fossae  Tholus        Patera
390°        300°        270°        240°    210°    180° 0°
Schioparelli      Schroeter   Fournier              Gale
Flaugergues                                  Herschel
Huygens                              HESPERIA    Graff
Millochau              PLANUM          Maadim Vallis
Bakhuysen      Terby              Molesworth
HELLAS          Arrhenius
-30°  Le Verrier      Hellespontus          Kepler    Cruls    -30°
Kaiser  Montes   PLANITIA     Tikhov      Tycho Brahe
Proctor          Secchi          Mendel
Barnard      Huxley
Holmes              Liais
-60°                                  -60°
South

S

a                                                    b

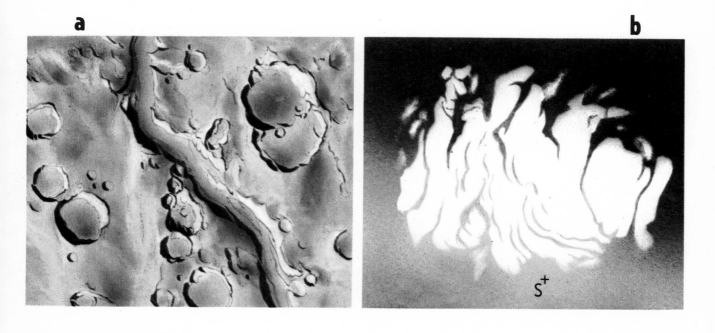

S⁺

# JUPITER
## GALILEAN
## MOONS

The first feature apparent to a casual observer of Jupiter using a small telescope is its string of moons — a 'solar system' in miniature. In 1610 Galileo Galilei discovered the four largest of Jupiter's moons. At the time of writing the number of discovered moons totals 14, and there are probably more yet to be found; but only the first 4 can be observed with a small telescope. Their name and diameters are:

I    — Io, 3,640 km
II   — Europa, 3,070 km
III  — Ganymede, 5,220 km
IV   — Kallisto, 4,890 km

The moons circle round Jupiter more or less in the plane of the ecliptic, so that they may be observed passing alternately from one side of the planet to the other and back, as depicted in Figure a, which shows the configuration of the Galilean moons on four consecutive days.

The following phenomena may be observed with a telescope (Figure b and c): 1 — western elongation, 2 — moon disappears behind the planetary disc, 3 — moon leaves occultation, 4 — moon enters umbra of Jupiter (beginning of lunar eclipse), 5 — moon leaves umbra of planet (end of eclipse), 6 — eastern elongation, 7 — moon enters disc, 8 — moon's shadow falls on disc of Jupiter, 9 — moon leaves disc, 10 — shadow of moon leaves disc.

In practice we usually observe only some of the above phenomena, depending on which moon it is and from which direction Jupiter is illuminated by the Sun. Forecasts of the respective phenomena may be found in astronomical year-books.

daily during period when Jupiter is visible — see astronomical year-book and the tables on pp. 161—175

a — motion of moons I—IV during four consecutive days
b — diagram of phenomena
c — satellite occultations and transits across the planet's disc

*1676 Römer (Denmark) — finite speed of light proved by observations of Jupiter's moons.*
*1965 Kalinyak (USSR) — discovery of an atmosphere on Europa, Ganymede and Io.*
*1974 Pioneer 11 (USA) — photographed the satellites Io, Ganymede and Kallisto.*

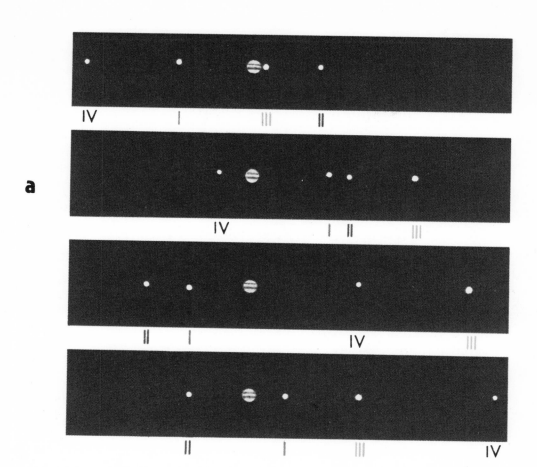

**a**

IV     I     III     II

IV     I   II     III

II     I     IV     III

II     I     III     IV

**b**

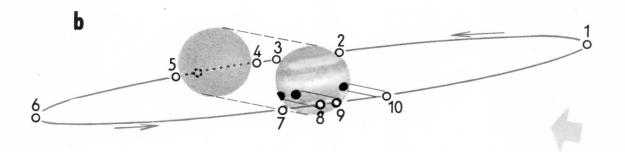

**c**

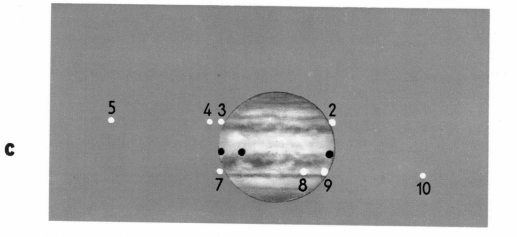

# JUPITER
# APPEARANCE
# OF THE PLANET

Jupiter seen through the telescope looks like a markedly flattened disc with dark and light belts parallel to the planet's equator. The belts are cloud formations in the dense atmosphere, which is composed predominantly of hydrogen and helium with an admixture of methane, ammonia and other compounds. Jupiter has an internal source of energy which affects circulation in the atmosphere — probably the planet is slowly shrinking under its own gravity, turning gravitational energy into heat. Rising currents of gas appear as light zones and descending currents are in the region of the dusky belts. The temperature on the top of the clouds is $-145°C$.

The number of belts, their width, intensity, division and coloration are continually changing (pictures at top and bottom). A very interesting feature is the oval-shaped Great Red Spot (GRS) with a diameter of more than 30,000 kilometres — big enough to swallow the Earth — which has been observed for several centuries. It is probably a stable eddy in the atmosphere, a giant jovian storm. The belts are categorised as follows:

*Dusky belts:* 2, 4, 6, — northern temperate belts, 8 — northern equatorial belt, 10 — equatorial belt, 12 — southern equatorial belt, 14, 16, 18 — southern temperate belts.

*Light zones:* 1 — northern polar region, 3, 5 northern temperate zone, 7 — northern tropical zone, 9, 11 — equatorial zone, 13 — southern tropical zone, 15, 17 — southern temperate zone, 19 — southern polar region.

The rotation period of zones 9, 10 and 11 (Rotation System I) is 9 hours 50 minutes, the other zones (Rotation System II) have a mean rotation period of 9 hours 56 minutes.

*1932 Wildt (USA) — discovery of methane and ammonium in the spectra of major planets.*
*1955 Burke and Franklin (USA) — discovery of electromagnetic radiation of radio frequency from Jupiter.*
*1973 Pioneer 10 (USA) — commencement of direct exploration of Jupiter.*

daily during period when Jupiter is visible — see astronomical year-book and the tables on pp. 161—175

Centre: marking of belts and zones on Jupiter

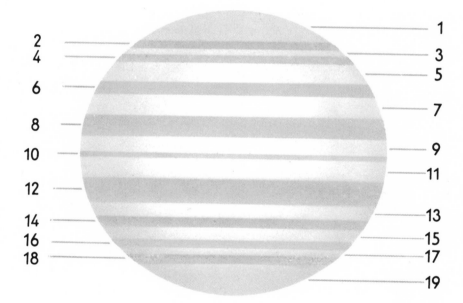

2
4
6
8
10
12
14
16
18

1
3
5
7
9
11
13
15
17
19

# SATURN
## VISIBILITY OF RINGS

In characteristics and appearance Saturn is very much like Jupiter, including the bands on the planetary disc. The surface temperature is −150°C. Saturn is encircled by 10 known satellites and rings.

Saturn's rings are composed of numerous small satellites — chunks of frozen methane, ice and other debris — each anything from a few centimetres to a few metres across. It is estimated that the overall thickness of the rings is several kilometres. The outer diameter of ring A is 278,000 kilometres. The outer, moderately bright ring A is separated from the main and very bright ring B by the dark Cassini division, which is about 5,000 kilometres wide. The transparent (crêpe) ring C looks like a darker band against the planetary disc. The faintest ring D, reaching to the surface of the planet, cannot be observed visually.

The appearance of the rings depends first and foremost on Saturn's position in its orbit round the Sun (picture at top). For example, in 1973 the rings were opened to the utmost and it was their southern aspect that was observed from the Earth. Since then the rings appear to be closing in, and in 1980 observers will be unable to see them for a brief period when they are edge-on to the Earth. In the following years they will show their northern aspect to the terrestrial observer. During observations note also the shadow cast by the planet on the rings, looking as if a piece has been bitten out of them.

daily during the period when Saturn is visible

top: visibility of the rings in 1973—2002
bottom: Saturn as it looks through the telescope; designation of the rings

1655—1659 *Huygens (Holland) — discovery of Saturn's rings and the satellite Titan.*
1894—1895 *Bielopolskii (Russia) and Keeler (USA) — spectroscopic proof that the rings are composed of meteoritic material.*
1948 *Kuiper (USA) — proof that the particles of which the rings are composed are coated with ice.*

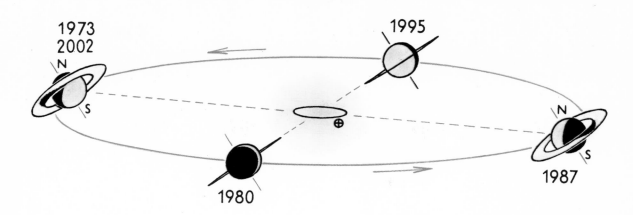

1973
2002

1995

1980

1987

A B C D ♄ D C B A

# URANUS, NEPTUNE AND PLUTO

Between 1975 and 1990 the last three superior planets will be moving on the constellation background of Virgo, Libra, Scorpius and Sagittarius (b).

*Uranus* is visible with binoculars and its disc can be distinguished with a telescope with about 100-fold magnification. Because of the extraordinary inclination of its rotational axis to the orbital plane, Uranus is tilted towards the Sun (and towards the Earth) alternately with its equatorial region (last in 1965) and one of the two poles (the North Pole will be inclined most in 1985). Details on the planet are visible only with a large telescope. In 1977 faint rings were discovered round Uranus; these, however, cannot be observed with equipment used by the amateur.

*Neptune* can be located with more powerful binoculars. Viewed through a small telescope it is indistinguishable from the stars. In order to locate Neptune it is necessary to have a detailed star map showing the position of the planet.

*Pluto* appears like a star of 14th magnitude in the sky and can therefore be observed only with larger, mainly photographic, instruments. On photographs made, for example, at intervals of several days, the planet will reveal itself by its motion among the stars.

From January 1979 to March 1999 Pluto will be closer to the Sun than Neptune. Pluto will be at perihelion (closest approach to the Sun) in September 1989.

Uranus and Neptune:

a detailed star map is a must — see astronomical yearbook

a — visibility of Uranus from 1923 to 2007
b — approximate positions of Uranus, Neptune and Pluto in 1975—1990

*1781 13 March — W. Herschel (England) — discovery of Uranus.*
*1846 23 September — Gall (Germany) discovered Neptune, whose existence had been previously predicted separately by Leverrier (France) and Adams (England).*
*1930 13 March — Tombaugh (USA) discovered Pluto.*

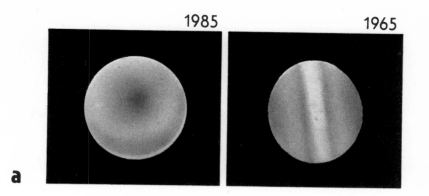

1985          1965

**a**

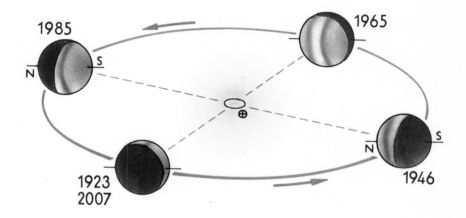

1985      1965
N   S

S

N    S

1923
2007        1946

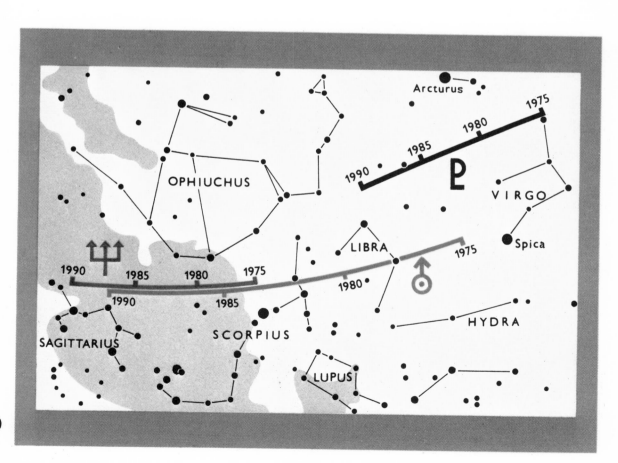

Arcturus

1975
1980
1985
1990

♇

VIRGO

Spica

OPHIUCHUS

LIBRA
1975
1980
♂
1990   1985   1980   1975
1990     1985

HYDRA

SAGITTARIUS    SCORPIUS

LUPUS

**b**

Circling round the Sun between the orbits of Mars and Jupiter are thousands of small bodies called minor planets, planetoids or asteroids — objects so small that they look like stars even through a telescope. Some asteroids can be seen with binoculars and one (Vesta) occasionally even with the unaided eye. The belt of asteroids (A) lies between the 2nd and 4th astronomical unit from the Sun with most of the orbits concentrated between 2.2 and 3.2 astronomical units (belt B). Besides these there are also asteroids with eccentric orbits (e.g. Adonis, Amor, Hidalgo — Figure a), some of which may come very close to the Earth. At least one asteroidal object, Chiron, has been found between the orbits of Uranus and Saturn, and there may be others in the same region. They cannot be observed with amateur equipment.

Asteroids are mostly observed by photographic means. Photographs made at various time intervals reveal the asteroid by its shift in position (b). A longer time exposure shows the motion of the asteroid (c).

Diagram d compares the dimensions of the Moon and five of the largest known asteroids. All told, there are some 110 asteroids which have a diameter of more than 100 kilometres. The typical asteroid, however, has a diameter of only 1 to 2 kilometres. Asteroids are objects of irregular shape, like Mars' two satellites Phobos and Deimos; their photographs, made by Mariner 9 and Viking space probes, give a good idea of what asteroids look like, and it is now thought that they are indeed asteroids themselves, captured by the gravitational pull of Mars.

 (telephoto lens),

for period of visibility and information on positions see astronomical year-book

a — orbits of minor planets
b, c — photographs showing the motion of minor planets
d — comparison of the largest asteroids with the Moon

*1801 1 January — Piazzi (Italy) — discovery of the first minor planet (Ceres).*
*1891 Wolf (Germany) — first photographic discovery of a minor planet.*

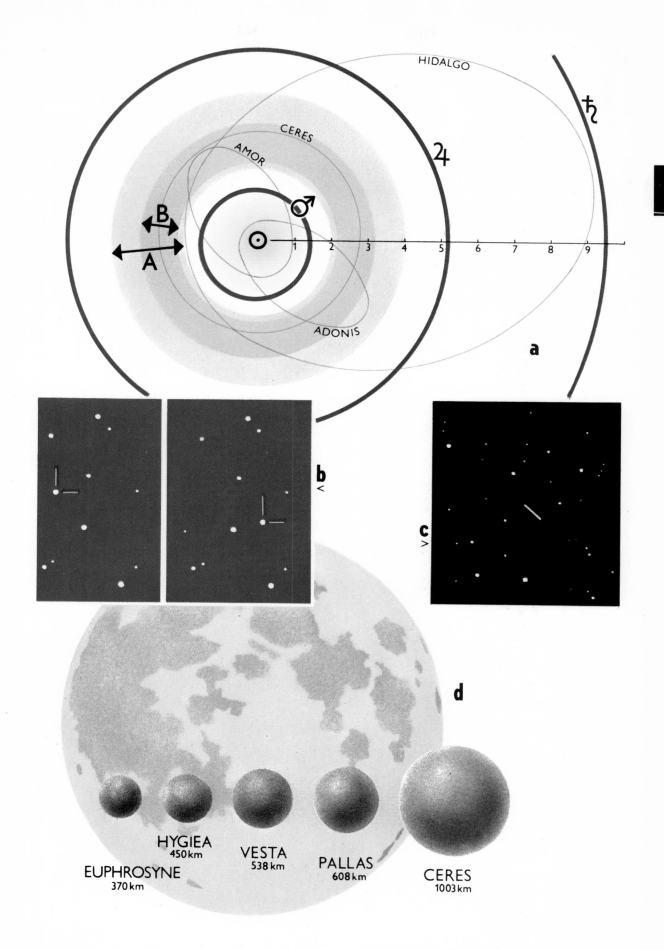

a

b

c

d

EUPHROSYNE
370 km

HYGIEA
450 km

VESTA
538 km

PALLAS
608 km

CERES
1003 km

HIDALGO

CERES

AMOR

ADONIS

A

B

A large number of meteoric bodies (meteoroids) daily enter the Earth's atmosphere, where they heat up to vaporization. The gas round the dying body flares up for a brief interval and it appears as a bright streak, a *meteor* (commonly known as a falling star).

Sporadic meteors flying in different directions arrive at the Earth all the time. On certain days of the year, however, meteors fall in greater numbers and appear to be coming from the same spot in the sky. This happens when the Earth passes through a meteor shower, a stream of meteoric bodies circling in an ellipse round the Sun. The paths of all the meteors in such a stream are parallel (c), but to the terrestrial observer they appear to shoot out fanwise from a single point — *radiant* R — because of the distortion caused by perspective (e.g. meteor M appears in the sky as M' — Figure b). Streams, or showers, are named after the constellation from which they appear to come, such as the Perseids, visible annually in the first half of August; the Geminids, visible in mid-December; and others. Bright meteors may be photographed even with an ordinary camera on a tripod with a long time-exposure; the stars will appear as parallel curves (a) in the resulting photograph, caused by the Earth's rotation, while the meteor trails appear as straight lines.

during the period of activity of the major meteoric streams — see astronomical yearbook

a — photograph of a meteor
b — shower meteors
  R — radiant
  O — observer
c — passage of the Earth through a meteoric stream

1866 Schiaparelli (Italy) -- proved that the orbit of the Perseids is identical with the orbit of the 1862 comet; proof of the relationship between meteoric showers and comets.
1885 Weinek (Bohemia) — first photographed a meteor.

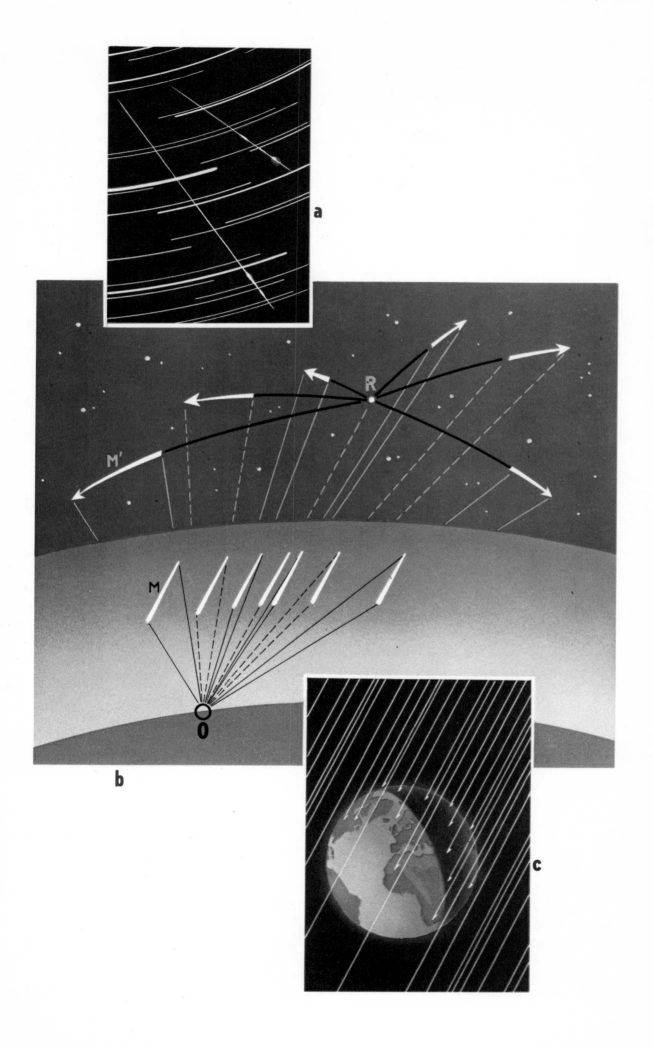

a

R

M'

M

O

b

c

# COURSE OF THE FALL
# OF A METEORITE

rare, chance phenomenon, very occasionally visible also in the daytime

left — passage of a bolide
right — phenomena occurring during a meteoric body's passage through the atmosphere

Some bright meteors explode in the sky, may flash briefly brighter than Venus, and are called *bolides.* The initial stage of such a spectacular light phenomenon is a larger meteoric body (of mass and weight about one kilogram or more). If you observe a bolide, record the phenomenon in as much detail as possible (time, duration of passage, path among the stars, position at the beginning and end of its trail, and so on) and send in a report to the nearest observatory or amateur astronomy group.

A meteoric body in space (a) is not visible from the Earth. On entering the atmosphere friction heats the body and makes it and the air along its path glow (b). This occurs at altitudes of more than 100 kilometres. In the denser layers atmospheric resistance increases considerably, the surface of the body becomes molten (at temperatures of 2,000 to 3,000° C) and a cushion of compressed air, which also glows, forms in front of it (c). Usually the whole body vaporizes and is dispersed about 40 kilometres or more above the ground. Bodies of sufficient size and density, however, penetrate deeper and are braked by atmospheric resistance at altitudes of about 20 to 30 kilometres with some material still not consumed. On reaching subsonic speed the air cushion detaches itself and changes into a sound wave which can sometimes be heard as a detonation (several minutes after the extinction of the bolide). The braked body (or its fragmented particles — d) rapidly cools and falls to the Earth. The recovered remnants of such a bolide (they are composed either of stone or iron) are called *meteorites* (e)

1794 *Chladni (Germany) — theory of the cosmic origin of meteorites.*
1803 *Shower of meteoric stones (2,000 pieces) in France; cosmic origin of meteorites proved.*
1959 *The 'Luhy' meteorite photographed in Czechoslovakia before impact — first exact determination of the orbit of a meteoric body in the Solar System.*

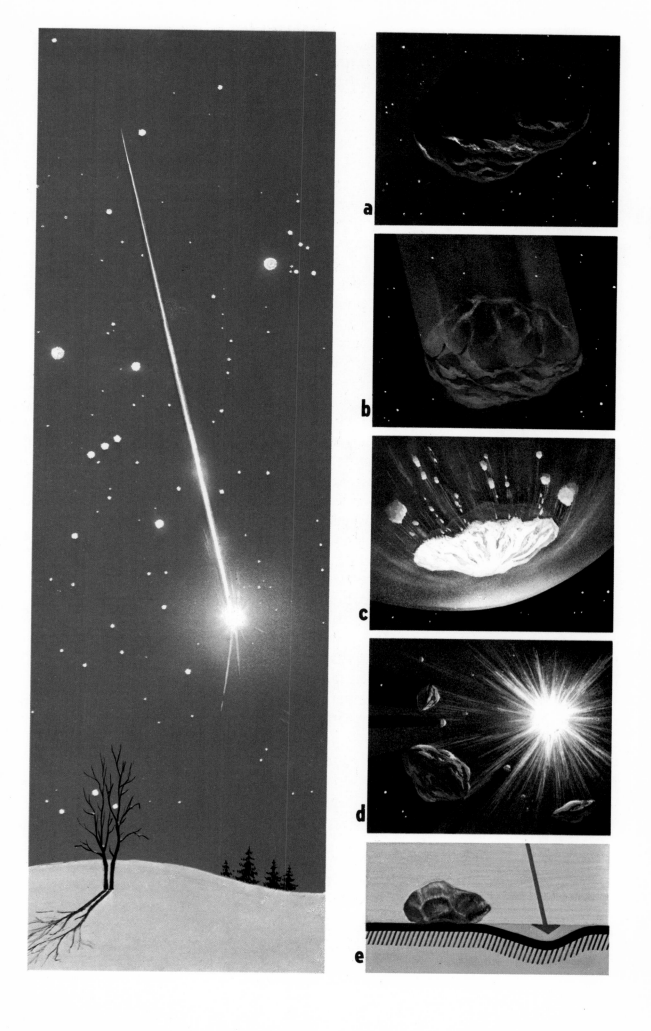

# COMETS
## ORBITS, DEVELOPMENT OF A TAIL

Comets revolve around the Sun in orbits which are very elongated, usually elliptic, diving in very close to the Sun before moving out again into space. Comets with an orbital period of less than 100 years are termed short-period and those with a period of more than 100 years long-period comets. The well-known Halley's comet has an orbital period of 76 years and moves away from the Sun to beyond the orbit of Neptune (a). In the case of comets with known orbits it is possible to forecast their return to the Sun; new comets appear unexpectedly.

At great distances from the Sun a comet is reduced to a solid *core* (b 1), composed of dust and larger meteoric particles frozen in ice of methane, ammonia, carbon dioxide and other compounds. With increasing proximity to the Sun this 'dirty snowball' becomes heated and begins to evaporate and surround itself with a gaseous *coma* (2), which together with the core forms the *head* of the comet (3). Near the Sun the liberated gases and dust particles are accelerated by its radiation pressure away from the Sun, streaming away to form a *tail* (4). The core, coma and tail combine to give the comet its overall appearance (5).

The tail of a comet is always turned away from the Sun (c) and is longest (in many instances tens of millions of kilometres) when the comet is at perihelion. The matter from the tail of a comet is irretrievably lost in space, so that after many orbits a comet gradually evaporates into insignificance.

rare phenomenon

a — Halley's comet: positions and speed in orbit (S — Sun)
b — parts of a comet
c — development of the tail

1705 *Halley (England) — discovered periodicity in the return of certain comets to the Sun. First prediction of the return of a comet.*
1812 *Olbers (Germany) — hypothesis that the tail of a comet is composed of particles expelled from the head of the comet by some force emanating from the Sun.*

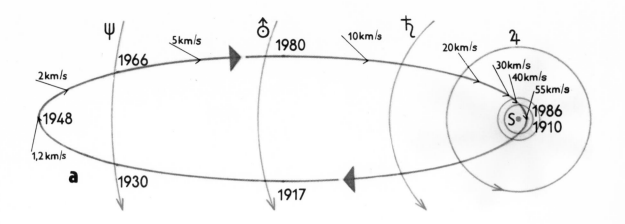

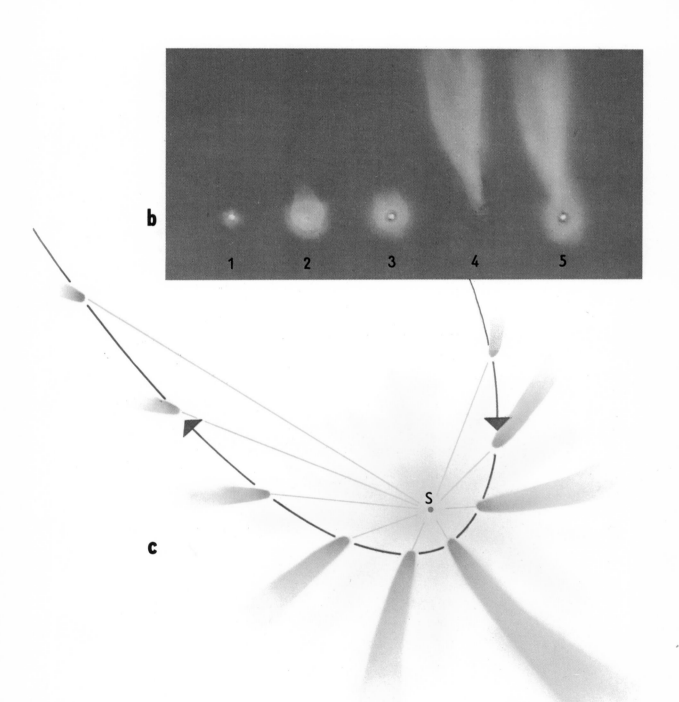

# TYPES
# OF COMETARY TAILS

Some 10 comets, including the expected (periodic) and newly discovered, are observed every year. Most, however, are objects so faint that they pass unnoticed. Through the telescope they look like hazy stars or nebulae (a).

Once in several years we can see one of the brighter comets with the unaided eye but only several times in a century can we enjoy the spectacle of a comet with a long conspicuous tail attracting the interest of onlookers and sometimes even causing a sensation. The brightest comets of the past decade include Ikeya-Seki (c) and West's (b) comets. Comets are named after their discoverers; these include many amateurs. The discovery of a new comet takes many hours of observation requiring both patience and perseverance.

The shapes of cometary tails are divided into four groups (picture at top):

$I_0$ — gaseous, straight tails, directed away from the Sun

$I$ — gaseous tails, slightly curved in the direction opposite to that of the comet's motion

$II$ — dust tails strongly curved in the direction opposite to that of the comet's motion

$II_0$ — straight dust tails markedly inclined away from the direction toward the Sun

Sometimes we observe a small so-called sunward-tail, directed straight towards the Sun.

during the period when the comet is at perihelion

at top — types of cometary tails

a — appearance of a faint comet

b — West's comet from 1976

c — Ikeya-Seki comet from 1965

*1881 Huggins (England) and Draper (USA) — first photographed the spectrum of a comet — proof that comets contain glowing gas and also solid particles.*

*1900—1910 Lebedev (Russia) — proof of the effect of radiation pressure on dust and gaseous particles.*

*1956 Discovery of the radio radiation of comets (the comet Arend-Roland).*

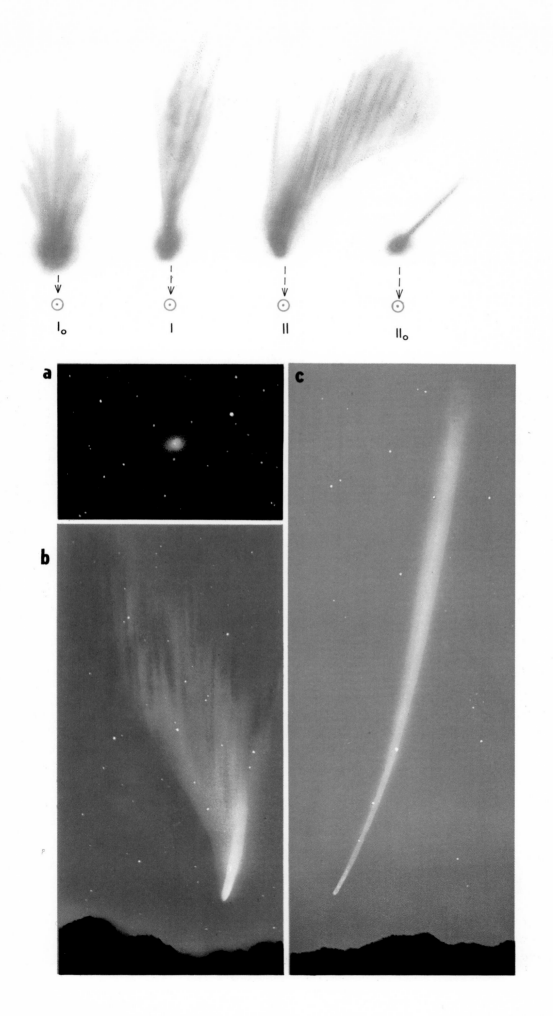

The Sun and other visible stars are members of a stellar system called the *Galaxy* (or Milky Way Galaxy). Our stellar island contains some 400 thousand million stars concentrated largely in spiral arms twisting round the nucleus of the Galaxy and forming a flattened disc. Located in this disc is our Sun, 8,500 parsecs distant from the centre of the Galaxy. The entire diameter of the disc is an estimated 30,000 parsecs (approximately 100,000 light years), so we are more than halfway out from the centre in something of a galactic backwater, marked with a circle opposite.

Stars and galactic formations with similar characteristics (motion, age) form subsystems in the Galaxy. The greatest mass is in the flat subsystem — the stellar disc, containing also a layer of gas and dust (a dark dividing belt visible to any beings who might be viewing the Galaxy from the side), dark and bright nebulae (b), open star clusters (c) and other, small objects. Planetary nebulae (d) are transient features. Farther from the disc are the globular star clusters (a), part of the spherical subsystem, forming round the centre of the Galaxy a halo about 40,000 parsecs in diameter (small white discs in the top picture).

Because there is so much material in the disc, in the plane of the Galaxy we can only see objects closer than 2,000 to 3,000 parsecs from the Sun. In other directions 'galactic windows' provide an unhindered view into distant space where we can observe other galaxies (p. 134).

Observation
see p. 112

top — schematic view of the Galaxy from the side
bottom — schematic view of the Galaxy from above
yellow ring — position of the Sun

1785 W. Herschel (England) — discovery of the motion of the Sun in relation to the surrounding stars.
1918 Shapley (USA) — first realistic model of our Galaxy; the Sun is located 30,000 light years from the centre of the Galaxy.
1927 Oort (Holland), Lindblad (Sweden) — discovery of the rotation of our Galaxy.

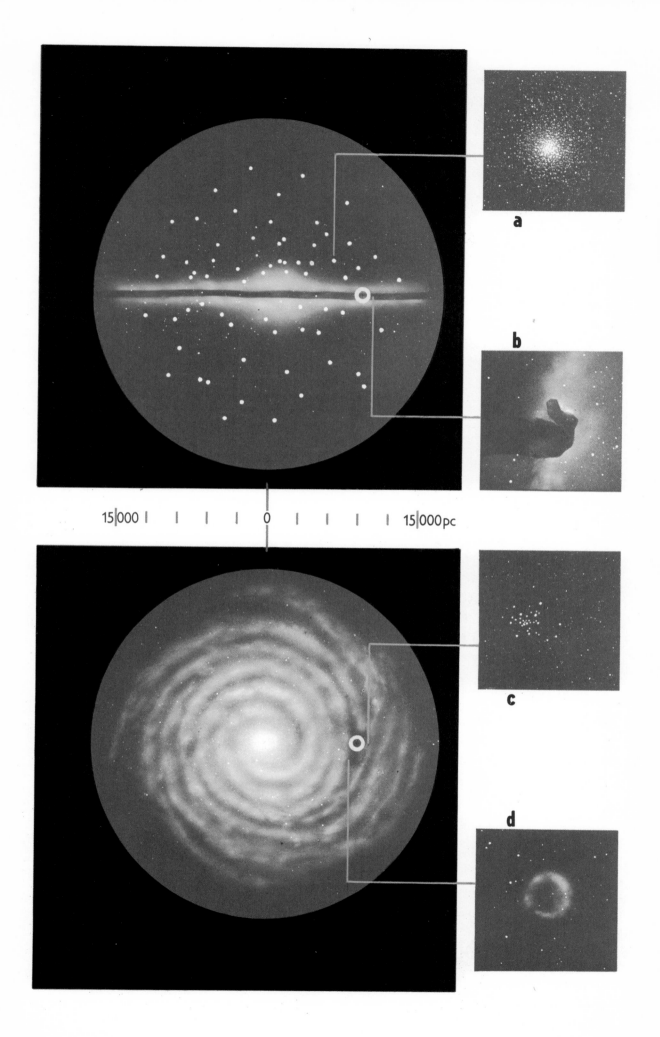

a

b

15 000 | | | | 0 | | | | 15 000 pc

c

d

The position of the Sun amidst the spiral arms some distance from the centre of the Galaxy (yellow ring in inset picture at right) makes it impossible for us to view the entire stellar island in which we live. If we were to look about us in the plane of the galactic disc, the neighbouring spiral arms would appear in the celestial sphere as a hazy belt of indefinite outline girdling the whole sky, both north and south. This belt is called the Milky Way; this term is sometimes used for our whole Galaxy.

The picture shows part of the Milky Way as it looks viewed with the unaided eye between the constellations Cassiopeia and Scorpius. The best time for observing this section is in summer and autumn. In the constellation of Cygnus (the Swan) the Milky Way is divided into two arms by a rift composed of dark clouds of interstellar matter which greatly limits our view of the Galaxy. The brightest parts of the Milky Way are in the region of the constellations Sagittarius (the Archer) and Scorpius (the Scorpion) because the centre of our Galaxy is located near Sagittarius.

The Milky Way contains numerous star clusters and nebulae which can be observed even with a small telescope.

 (with parallactic mounting)

daily, best of all in summer and autumn

Cygnus — the Swan
Aquila — the Eagle
Sagittarius — the Archer
Scorpius — the Scorpion
see also maps on p. 153

1785 W. Herschel (England) — study of the structure of the Milky Way; conception of a flattened stellar system based on observations.
1889 Barnard (USA) — first photographed the Milky Way.
1951 First radio-astronomical observation of the spiral structure of the Galaxy.

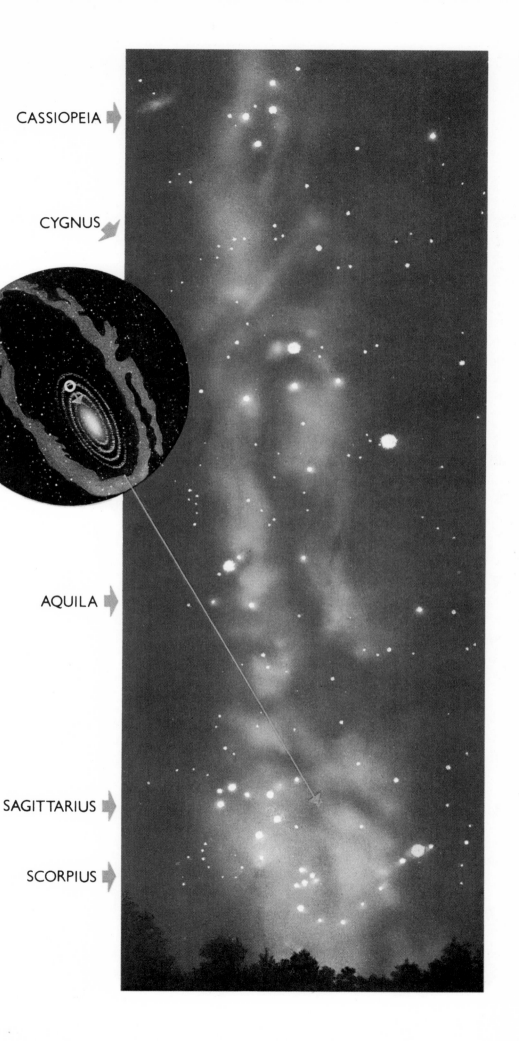

CASSIOPEIA

CYGNUS

AQUILA

SAGITTARIUS

SCORPIUS

# CHARACTERISTICS
## OF THE STARS
## BRIGHTNESS,
## DISTANCE, DIAMETER

What, in the star-studded sky, is illusion and what is real?

The constellations are the product of man's imagination. Individual stars orbit round the centre of the Galaxy in the same direction but at different speeds. But superimposed on these circular motions are variations which show up, for example, in the changes in the image formed by the five brightest stars of Cassiopeia from the year 50,000 B. C. to 1980 to 50,000 A. D. (at top).

The present arrangement of the five stars in the shape of a W is the result of the projection of a certain spatial configuration of the stars on the celestial sphere (at bottom). The relation of *apparent stellar magnitudes* m (top map m) is quite different from that of the stars' actual luminosity, expressed as *absolute stellar magnitude* M (magnitude M is denoted by the diameters of the discs in diagram D and map M). The brightest in absolute terms is star gamma; beta is about 200 times fainter. Gamma, however, is 14 times farther away than beta and therefore both appear to be equally bright.

If we compare the actual diameters of the stars R (map R) with their luminosity we see that the brightest is not the largest star (alpha — 42 times larger than the Sun, surface temperature 4,100 K) but the hottest (gamma, with a temperature of 22,000 K).

m — the stellar discs on the map correspond to apparent stellar magnitude m
D — distance
l. y. — light years
M — the discs correspond to absolute stellar magnitude M
(The Sun is shown at bottom left)
R — actual diameters of the stars in comparison with the Sun (at bottom left)

*1718 Halley (England) — discovery of the proper motion of stars.*
*1835—1840 Struve (Russia), Bessel (Germany) and Henderson (England) — first measurements of stellar distances.*
*1937—1939 Bethe (USA) — theory of thermonuclear reactions in the stellar interior as the source of stellar energy.*

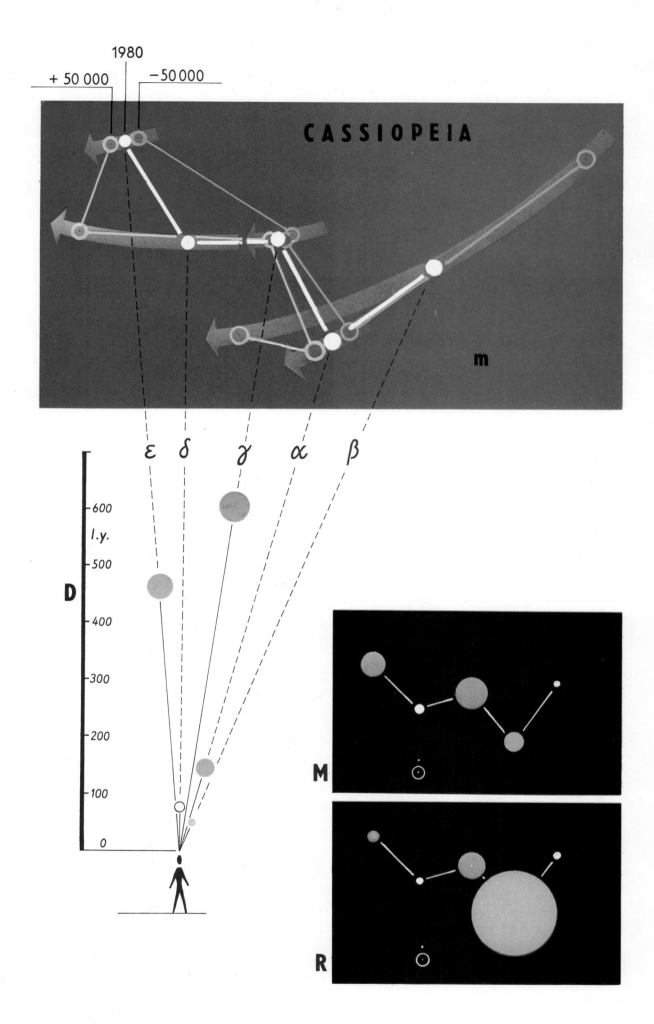

CASSIOPEIA

+ 50 000    1980    − 50 000

m

ε  δ  γ  α  β

D

l.y.
- 600
- 500
- 400
- 300
- 200
- 100
- 0

M

R

# COLOUR OF STARS
## ENERGY
## DISTRIBUTION
## IN THE SPECTRUM
## OF A STAR

Stars have different colours. Some are striking — the orange coloration of Aldebaran in Taurus and Arcturus in Boötes, the reddish colour of the star Antares in Scorpius, and those listed in the accompanying plate.

The colour of a star is closely correlated with its surface temperature. Relatively cool stars with temperatures of about 3,000 K (Kelvin) appear to be red; yellow stars such as our Sun have a temperature of about 6,000 K; and white to blue-white stars have a surface temperature as high as 10,000 K and more.

The apparent coloration of stars is caused by the distribution of radiant energy in the spectrum. As temperature rises, the maximum radiant energy shifts to the shorter wavelengths and the luminosity of the star also increases (denoted in the graphs by the height of the maximum). For example, a star with surface temperature of 3,000 to 4,000 K radiates its greatest amount of energy in the region of red light and infrared (IR) radiation; thus such a star appears to be coloured red.

The surface temperatures of stars are determined on the basis of their spectrum. Stars are divided according to their temperature into several so-called spectral classes denoted by the capital letters O, B, A, F, G, K, M. Classification is based on the fine structure of the spectrum, particularly the presence of spectral lines corresponding to certain atoms.

Colour of stars:

 (very sensitive colour film)

top:
diagram
    showing the decomposition of light into a spectrum
bottom:
UV — ultraviolet radiation
IR — infrared radiation
nm — wavelength in nanometres
K — surface temperature of star, temperature in kelvins
O, B, A, F, G, K, M — spectral classes

*1859—1864 Kirchhoff (Germany) — discovery of fundamental laws important for spectral analysis.*
*1872 Draper (USA) — first photographed a stellar spectrum.*
*1913 Bohr (Denmark) — theory of the structure of the atom permitting the interpretation of the series of lines in stellar spectra.*

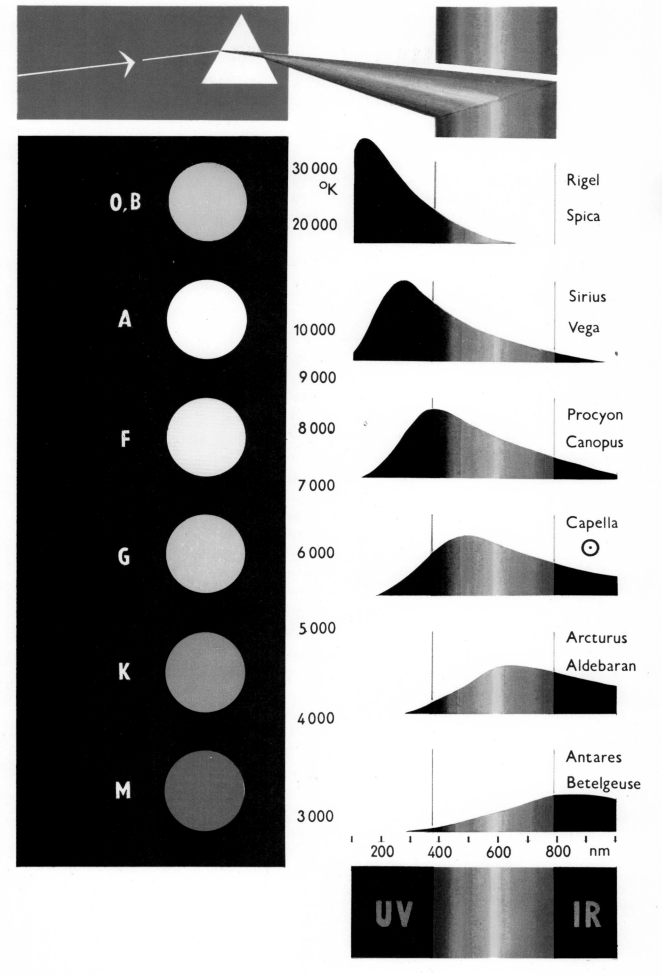

O, B

A

F

G

K

M

30 000 °K

20 000

10 000

9 000

8 000

7 000

6 000

5 000

4 000

3 000

Rigel

Spica

Sirius

Vega

Procyon

Canopus

Capella

⊙

Arcturus

Aldebaran

Antares

Betelgeuse

200    400    600    800    nm

UV                              IR

# TEMPERATURE — LUMINOSITY DIAGRAM GIANTS AND DWARFS

The characteristics of stars and their evolution may be tracked graphically on the temperature-luminosity diagram, also known as the Hertzsprung-Russel or H-R diagram. If we plot the stars in this diagram according to their temperature (spectral class) and luminosity (or absolute stellar magnitude M), they will not be scattered randomly but will cluster round certain lines (branches).

The greatest number is located on the *main sequence* a; our Sun is one of them. A large number of stars is located on the *giant* branch (g), fewer in the group of *supergiants* (c). The number of so-called *white dwarfs* (d) and *red dwarfs* (b) that we can observe is small, even though in actual fact red dwarfs form the majority of the stellar population. This is because such stars are very faint and difficult to observe. The diagram shows, for example, that class M stars may exhibit marked differences in luminosity: from 0.0001 (and less) to 10,000 times that of the Sun. This is caused by the great differences in their dimensions.

Shown in the top part of the plate are the relative dimensions of certain giants and supergiants in comparison with the small disc of the Sun. In the enlarged scale at the bottom, on the other hand, the yellow Sun stands out conspicuously, thus emphasizing the relatively small dimensions of the red and particularly white dwarfs in comparison with the Earth (brown disc).

a — main sequence
b — red dwarfs
g — giants
c — super giants
d — white dwarfs
M — absolute stellar magnitude
L:L☉ — ratio of the luminosity of a star to the luminosity of the Sun
Proxima Centauri — star closest to the Sun (it is the faintest component of the triple star α Centauri)

*1905 Hertzsprung (Denmark) — discovered differences in the luminosity of red stars and suggested their division into 'giants' and 'dwarfs'.*
*1911—1913 Russel (USA) and Hertzsprung compiled the spectrum-luminosity diagram.*
*1914 Adams and Kohlschütter (USA) — method of determining absolute stellar magnitudes (and indirectly also distances) from stellar spectra.*

Antares

Rigel

Arcturus

Capella

Sirius A

M

-5

10 000

c

$\frac{L}{L_\odot}$

0

100

g

a

+5

1

+10

0,01

d

b

+15

0,0001

O  B  A  F  G  K  M

20 000    10 000    8000    6000    5000    4000    3 000°K

Wolf 457

Van
Maanen

Sirius B

Proxima
Centauri

Wolf 359

# DOUBLE STARS (BINARIES)
## COLOURED BINARIES FOR TELESCOPIC OBSERVATION

Stars are usually not solitary occupants of space but form systems, the simplest of which are the double stars or binaries. The components of a double star each revolve around a common centre of gravity along a different elliptical orbit.

In a *visual double* the two components may be resolved with a telescope. A *spectroscopic double* looks like a single star even through the largest of telescopes and the fact that it consists of two components is revealed only in the spectrum. For the amateur, observation of a double star with a known angular separation of the two components (listed in a standard reference book) is an easy and reliable way of determining the actual resolving power of a telescope (p. 177) and thus testing its quality.

Double stars in which the two components are of different spectral classes and have strikingly different colours are among the beautiful sights which make even the most casual observation worthwhile. An interesting physiological phenomenon occurs when observing such stars — the retina of the observer's eye is overwhelmed by the light of the brighter component whose colour (say orange) predominates; the fainter component is then seen in a complementary colour (blue is complementary to orange). The plate at right shows some of the loveliest of coloured binaries (their colours in the picture are brighter than in reality). The system epsilon Lyrae is a quadruple star, made up of two binary pairs.

daily — choice according to picture at right, star maps and catalogues

Beside each double star is printed its name, apparent stellar magnitude of the two components and their distance in seconds of arc.

1782 *W. Herschel (England) — discovery of visual binaries, determination of orbital motion, first catalogue of (269) double stars.*
1889 *Discovery of spectroscopic binaries (several observers in the USA and Germany).*
1932 *Aitken (USA) — catalogue of 17,180 double stars.*

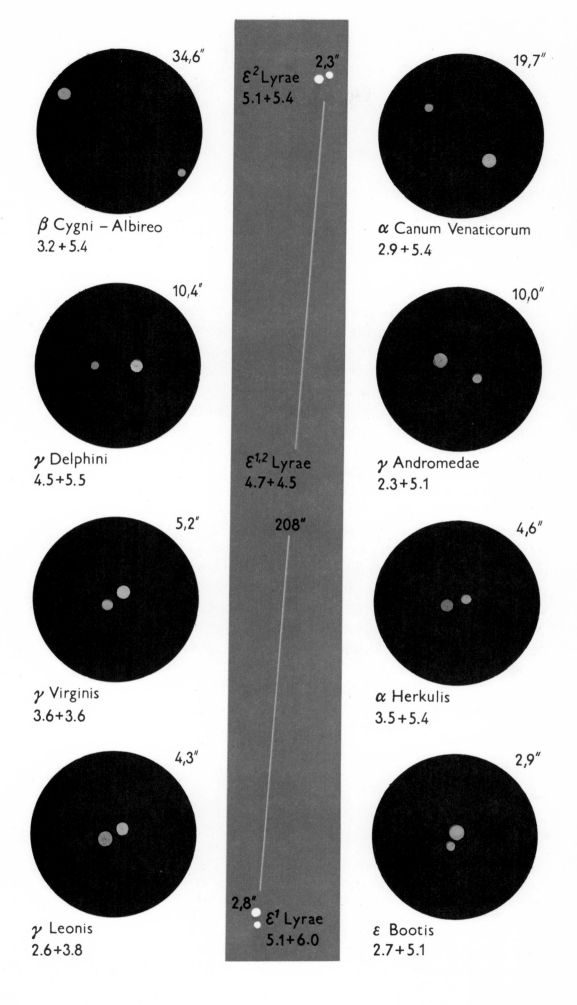

34,6"

β Cygni – Albireo
3.2 + 5.4

10,4"

γ Delphini
4.5 + 5.5

5,2"

γ Virginis
3.6 + 3.6

4,3"

γ Leonis
2.6 + 3.8

2,3"

ε² Lyrae
5.1 + 5.4

ε^{1,2} Lyrae
4.7 + 4.5

208"

2,8"

ε¹ Lyrae
5.1 + 6.0

19,7"

α Canum Venaticorum
2.9 + 5.4

10,0"

γ Andromedae
2.3 + 5.1

4,6"

α Herkulis
3.5 + 5.4

2,9"

ε Bootis
2.7 + 5.1

Stars whose brightness varies at regular or irregular intervals are, logically enough, called variables. The reasons for this variability may be geometric (as when the stars in a binary system eclipse one another) or physical (changes in the diameter and temperature of the star).

Beta Persei — Algol — is a representative example of an *eclipsing variable*. The fainter, cool star with diameter 3.8 times that of the Sun regularly eclipses the brighter, hot component with diameter 3.6 times that of the Sun, producing variations in over a period of 68.8 hours. The brightness of Algol changes from a maximum of 2.2$^m$ to minimum of 3.5$^m$, as shown in the accompanying *light curve*.

There are many types of physically variable stars, among the best known being the *cepheids,* so called after their prototype, delta Cephei. That star's diameter changes regularly in periods of 5.37 days (from 31.6 to 35.5 times the diameter of the Sun) and so does its surface temperature (from 5,300 K to 6,700 K) thus changing the apparent stellar magnitude from 3.6$^m$ to 4.3$^m$. The whole star literally pulses in and out like a balloon being inflated and deflated.

Variable stars also include novae and supernovae, whose sudden outburst (explosion) causes a great increase in brightness. This, however, is an altogether different phenomenon from regular variation.

during periods of the variable star's minimum — see astronomical year-book

1 — minimum brightness of Algol
2 — maximum
3 — secondary minimum
A — smallest diameter of delta Cephei
B — maximum brightness and temperature
C — largest diameter
D — minimum brightness and temperature

*1667 Montanari (Holland) discovered the variability of Algol.*

*1782 Goodricke (England) — assumption that the variability of Algol is caused by periodic occultation of the star by a dark body.*

*1880 Pickering (USA) determined the relative dimensions of the Algol system from analysis of the changes in its brightness.*

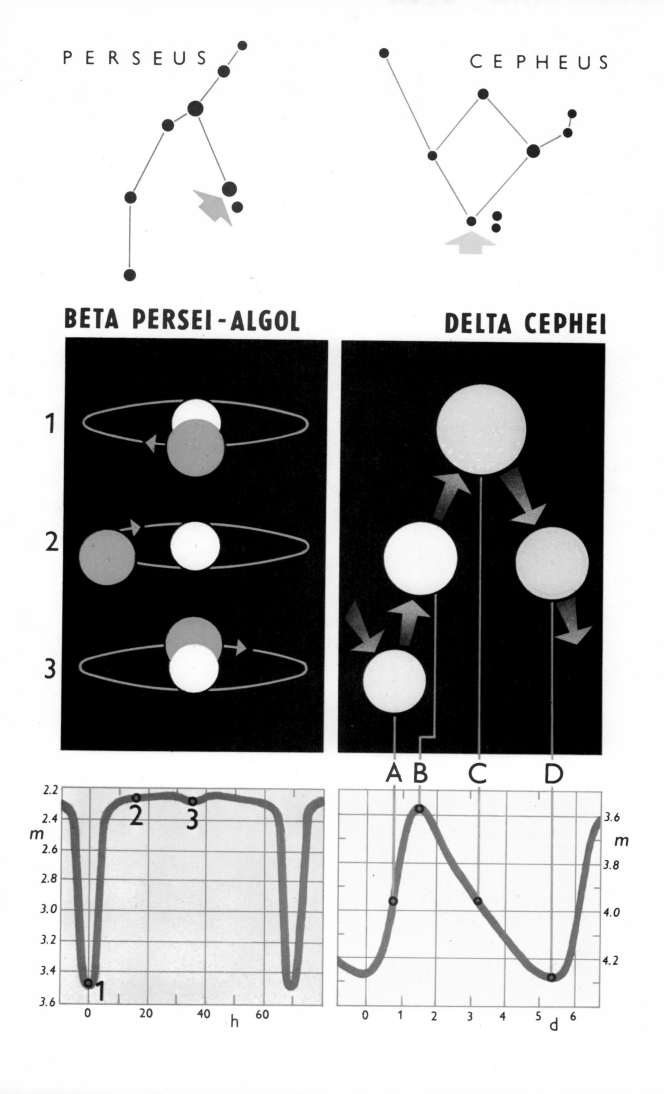

PERSEUS

CEPHEUS

BETA PERSEI - ALGOL

DELTA CEPHEI

1

2

3

A  B          C          D

# PLEIADES
## OPEN
## STAR CLUSTER

The Pleiades in the constellation of Taurus (the Bull) is a well-known open star cluster in which many of the individual stars can be distinguished with the unaided eye. Under ideal conditions we can see all 9 of the named stars, on rare occasions further 2 as well. From a lighted city an observer with good eyesight can expect to see 5 or 6 stars. The visibility of the Pleiades is a good test of the conditions of observation and quality of sight. For observation with binoculars or small telescope the plate includes a small map with notation of apparent stellar magnitudes.

The star cluster actually contains some 130 stars which were formed as a group about 60 million years ago and now, as a group, travel through space in the same direction (a so-called moving star cluster). This stellar system is some 30 light years in diameter and 410 light years distant.

The brightest stars in the Pleiades — white giants of spectral class B — rotate rapidly round their axis, up to a hundred times faster than the Sun. The cluster contains clouds of dust that appear as so-called reflection nebulae near the brightest stars, whose light they scatter. Under ideal conditions these nebulae are faintly visible even visually.

daily from late summer till early spring

top — maps of the brightest stars in the Pleiades and the star cluster as it looks on a photograph
bottom — maps of the Pleiades (numbers denoting apparent stellar magnitudes).

*1610 Galileo counted 40 stars in the Pleiades with his telescope.*
*1885 the Henry brothers (France) — first photographed the Pleiades: 1,421 stars, discovery of nebulae in the Pleiades.*
*1912 Slipher (USA) found that the spectrum of the nebulae in the Pleiades is similar to the spectra of the stars surrounded by the nebulae (proof of the existence of reflection nebulae).*

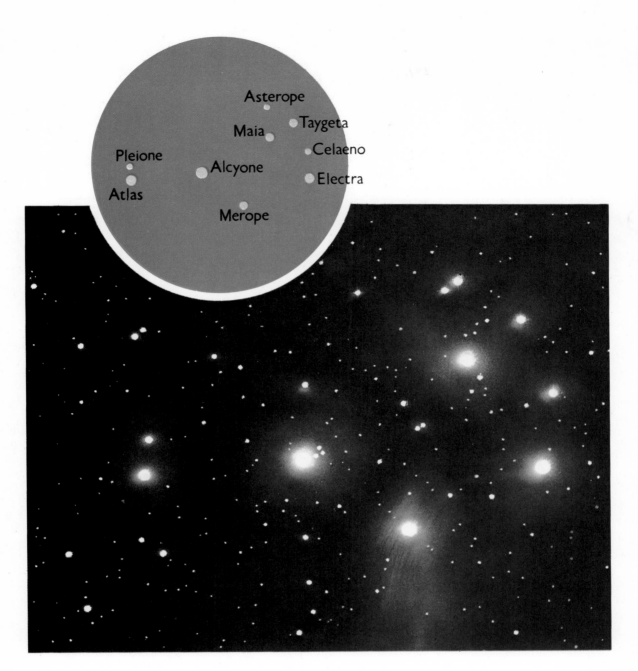

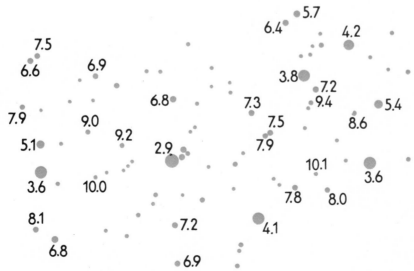

# STAR CLUSTERS
## OPEN
## AND GLOBULAR
## STAR CLUSTERS

Star clusters rank among the most popular objects of observation by amateurs. *Open star clusters* contain tens, hundreds and even thousands of stars bound to each other by the force of gravity. There are some several hundred such known clusters in our Galaxy, among the best-known being the double open cluster χ and h in the constellation of Perseus, visible with the unaided eye and some 8,000 light years distant. The M 44 — Praesepe-cluster (Beehive) in the constellation of Cancer (the Crab) can also be seen without a telescope as a hazy cloud larger than the Moon's disc. M 44 is 500 light years distant and contains some 60 stars.

Open star clusters are the youngest and *globular star clusters* the oldest formations in the Galaxy. The latter are conglomerations of spherical shape containing hundreds of thousands to millions of stars with a marked concentration of these toward the centre. They measure up to several hundred light years in diameter and form a halo 40,000 parsecs in diameter round the centre of our Galaxy.

The loveliest globular star cluster in the northern sky is M 13 in Hercules, visible even with the unaided eye. Seen through a small telescope it looks like a hazy cloud; larger instruments resolve it into a myriad of stars of 10th to 11th magnitude forming a halo in the centre. M 13 is 25,000 light years distant and measures more than 100 light years in diameter.

(long exposure!)

daily, choice according to star maps

top — χ and h Persei and M 44 as seen through a telescope
bottom — M 13 as it appears on a photograph

*1914—1918 Shapley (USA) found that globular star clusters form a cloud round the centre of the Galaxy — conception of the structure of the Galaxy made more precise.*
*1947—1948 Ambarcumyan (USSR) — discovery of stellar associations, groups of hot young stars — proof that stars are being formed even at the present time.*

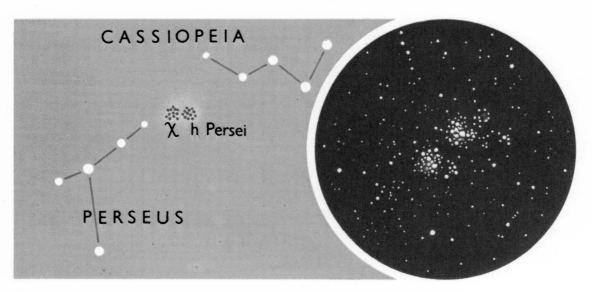

CASSIOPEIA

χ h Persei

PERSEUS

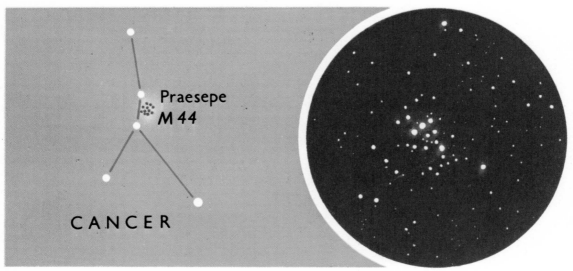

Praesepe
M 44

CANCER

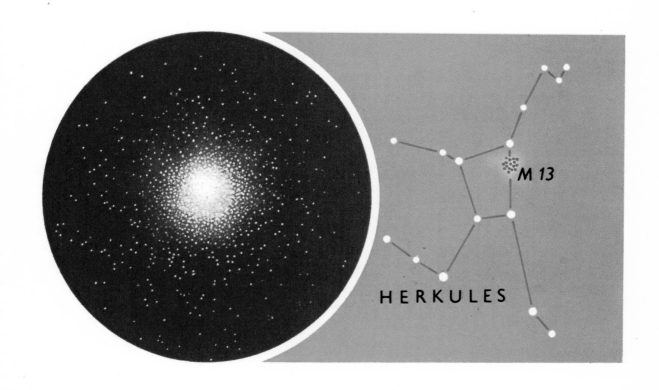

M 13

HERKULES

These have nothing in common with the planets, except that in some cases the object looks like a small disc. In actual fact they are glowing gaseous envelopes, usually spherical in shape, which expand at a speed of 10 to 50 km/s from a very hot central star (temperature up to 100,000 K). There are more than 1,000 known nebulae of this type.

M 57 — Ring Nebula in Lyra, first to be designated as a planetary nebula (Darquier, 1779). Distance 1,400 light years, actual size 0.6 × 0.8 light years. Central star (14.7$^m$) has a surface temperature of 75,000 K.

M 27 — Dumb-bell in the constellation of Vulpecula (the Fox) was discovered by Messier in 1764 and is the oldest known planetary nebula. It is bright, large and a good object for visual observation. Distance 975 light years, size 1.2 × 2.3 light years. Central star 13.4$^m$ with surface temperature of 85,000 K.

NGC 7293 — Helix in Aquarius (the Water Carrier) has an angular diameter equal to half the Full Moon but is very faint and difficult to observe visually; details are revealed only on photographs.

 (long exposures)

daily, choice according to star maps

the drawings at right show what the given nebulae look like in photographs

*1864 Huggins (England) — proved, on the basis of the spectrum of the planetary nebula in the Dragon, that some nebulae are composed of gases (it was previously believed that they were composed of stars).*

*1917 Campbell, Moore, Wilson (USA) — observation that planetary nebulae are in fact the expanding atmosphere of a central star.*

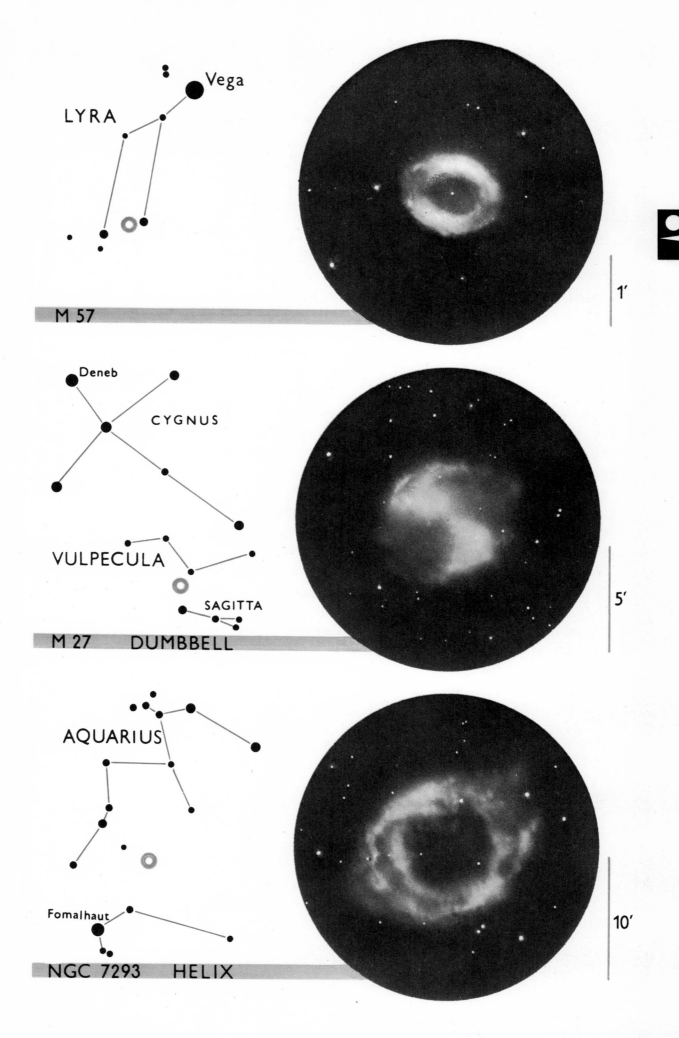

LYRA
Vega

M 57

1'

Deneb
CYGNUS
VULPECULA
SAGITTA
M 27  DUMBBELL

5'

AQUARIUS
Fomalhaut
NGC 7293  HELIX

10'

# HORSEHEAD
# NEBULA

Interstellar space is filled with a thin spread matter which is more concentrated in the region of the spiral arms (p. 110) and towards the centre of the Galaxy, where it forms large clouds. Interstellar matter consists mostly (99%) of gas (chiefly hydrogen), with interstellar dust forming the remainder. If a cloud of interstellar matter is illuminated by a nearby star, it is visible as a *diffuse nebula.* This may be a *dust* or *reflection nebula* if it shines only due to reflection of light by the dust particles in the cloud, or else a *gaseous* or *emission nebula* if intense ultraviolet radiation from a nearby, very hot, star ionizes the gas and causes the cloud's gaseous component to glow like the gas in a neon tube light. A combination of both processes gives rise to a gas-and-dust nebula.

Apart from rare exceptions, such as the great nebula in Orion (p. 133), difuse nebulae are very difficult objects for visual observation and they appear in their full beauty only on long-exposure photographs. The accompanying plate shows the diffuse gaseous nebula IC 434 in Orion with a cloud of dark interstellar matter (dark nebula) the shape of a horse's head extending out in front. This nebula is 1,100 light years distant.

 (rare exceptions)

 (long exposures)

daily, choice according to star maps

the IC 434 nebula with dark Horsehead Nebula as it appears on a photograph

1847 *V. Struve (Russia) — assumption that the light of stars is absorbed in the interstellar medium.*
1930 *Trumpler (USA) — proof of the existence of interstellar absorption of light.*

ORION

IC 434

M 42

# GREAT NEBULA
# IN ORION

The M 42 nebula in Orion may rightfully be described in superlatives as: the largest, brightest, loveliest and best object of observation for the amateur astronomer. It is visible even to the unaided eye and with a telescope it is possible to distinguish many details.

The drawing of M 42 (based on observations by G. P. Bond) contains all the details that can be registered visually. The brightest part of the nebula was named Regio Huygeniana by J. Herschel in 1826. Extending into this region is a dark bay — Sinus Magnus — at the end of which is a small striking polygon known as the Trapezium in Orion, composed of small 6th to 8th magnitude stars. Visually the angular diameter of M 42 is about 20′.

Details that were formerly laboriously sought visually for years are now obtained by a modern photographic telescope within minutes. Long exposures make it possible to capture even the outer and fainter parts of the nebula.

The true diameter of M 42 is about 20 light years, its distance 1,300 light years. It is a diffuse gaseous nebula whose visible radiation is excited by the UV radiation of a close, hot star. The Trapezium is composed of very young stars formed in the nebula about 10,000 years ago along with other stars, which, however, have now moved far away from their birth site.

 (various exposures)

daily from autumn to early spring

top — drawing based on visual observations by G. P. Bond in 1858—1864
bottom — M 42 as it appears on a photograph

1610 *Peirescius (France) discovered the great nebula in Orion.*
1656 *Huygens (Holland) — first detailed description of the M 42 nebula.*
1880 *Draper (USA) — first photographed M 42.*
1882 *Huggins (England) — first photographed the spectrum of M 42.*

# TYPES
# OF GALAXIES

Galaxies are other 'island universes' like our own Milky Way Galaxy in its entirety, but so far away that they, too, appear as faint nebulae in our sky. Galaxies differ in appearance, dimensions, proportion of stars of various types, total mass, quantity of interstellar matter they contain, and so on. In general, the sizes of galaxies range from several thousand to a hundred thousand light years across and the number of stars they contain may be several hundred thousand million. In 1975 the most distant of the known galaxies was 3 C 123, which is 8 thousand million light years away and can only be observed with the aid of modern professional equipment.

Galaxies are divided according to their appearance into elliptical galaxies, spiral galaxies, barred spirals and irregular galaxies. The spiral structure of some galaxies was discovered in the mid-nineteenth century by Lord Rosse. Visually galaxies look like hazy clouds or hazy stars; the structure is evident only on rare occasions. Details can be captured only with large photographic telescopes.

The Small Magellanic Cloud (SMC) is a companion of our Galaxy. This irregular galaxy is 220,000 light years distant and has an apparent diameter of 5° 20'. It is visible to the unaided eye as a 'severed fragment' of the Milky Way (see p. 157).

The Whirlpool galaxy M 51 in Canes Venatici (the Hunting Dogs) is 37 million light years distant and has an apparent diameter of 10'. It is linked by one arm with a smaller galaxy. The spiral structure is visible with a larger telescope.

1845 *Lord Rosse (Ireland) — discovery of the spiral structure of certain nebulae.*
1924 *Hubble (USA) — discovery that the 'spiral nebulae' M 31 and M 33 are composed of stars and that they are stellar systems similar to our Galaxy.*
1951 *Baade (USA) discovered an error in the original method of determining the distance of galaxies (according to cepheids): distance must be approximately doubled.*

(long exposures)

daily — choice according to star maps

at left:
top — elliptical galaxy
centre — barred spiral
bottom — irregular galaxy (SMC)
at right:
top — spiral galaxy M 51 in CVn
bottom — spiral galaxy (side view) according to the photographs.

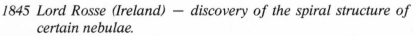

# ANDROMEDA GALAXY

The most distant object visible to the unaided eye is the galaxy in Andromeda (M 31), also known as the Andromeda Nebula. It is 2,200,000 light years distant and with its two companions M 32 and NGC 205 belongs to the Local Group of Galaxies, of which our Milky Way Galaxy is also a member. The true diameter of M 31 is more than 150,000 light years (apparent diameter approximately 3°).

Only the brightest central region of the Andromeda galaxy may be observed visually. Messier (1764) compared it to two pyramids placed base to base; the outer peaks of the pyramids are 40′ distant and their joint base is about 15′ long. The brightness increases towards the centre of the small hazy cloud.

The spiral arms are clearly visible on photographs made with even small telescopes and with the largest telescopes it is possible to distinguish in galaxy M 31 single stars, including variable stars, star clusters and diffuse nebulae. The structure of our Galaxy and that of M 31 are very much alike, thus giving us a good idea of what our stellar island would look like viewed from afar.

In the neighbouring constellation Triangulum (the Triangle) it is possible to observe another galaxy M 33 through a telescope with small magnification. It, too, belongs to the local system (distance 2.3 million light years).

 (small magnification)

 (long exposure)

daily from summer until the end of winter

Galaxy M 31 as it appears on a photograph
bottom: location of galaxies M 31 and M 33

1612 Marius (Germany) — first telescopic observation.
1742 Le Gentil (France) — discovery of the companion M 32.
1925 Hubble (USA) — determination of the distance of M 31 based on the observation of cepheids.

M 31

A N D R O M E D A

TRIANGULUM

M 33

ARIES

P E G A S U S

# EXPANSION
# OF THE UNIVERSE
# HUBBLE'S LAW

In the spectra of the galaxies one may observe an unusual phenomenon: the spectral lines are shifted to the red end of the spectrum and the farther the galaxy is from the Earth the greater this so-called *red shift.*

If the spectrum of a distant galaxy is photographed together with a well-known companion spectrum (in the accompanying picture it is the spectrum of helium), it is possible to measure the extent of the red shift. For example, the pronounced calcium lines have a wave length of 393.4 nm (line K) and 396.8 nm (line H) when photographed in the laboratory; in the spectra of four selected galaxies, however, this pair of lines is shifted tens and hundreds of nanometres towards the longer wavelengths.

Lines shift to the red end of the spectrum when the source of radiation is moving away from the observer (Doppler effect), and the faster it recedes the greater the red shift in the spectra of galaxies: galaxies are moving away from our Galaxy (except for some 20 members of the local system) at a speed that is greater for more distant galaxies.

The relation between the velocity $V$ and distance $r$ of the galaxy is given by the equation $V = H . r$ (Hubble's law), where $H$ is Hubble's constant. The value $H$ is not reliably known as yet; the data in the illustration correspond to $H = 50$ km/sec per megaparsec.

Our Galaxy does not occupy a special position in the Universe. If we could visit any of the other galaxies, we would observe the same phenomenon — that the other galaxies are moving away from us in all directions. We are part of an expanding universe where all the galaxies are moving away from each other. According to the big bang theory, the beginning of the expansion of our Universe as we know it was the gigantic explosion of an exceptionally hot and dense 'pre-atom' which occurred some 15—20 thousand million years ago.

1842 Doppler (Austria) — *discovery of the optical principle explaining the change in wavelength caused by the motion of the source of radiation in relation to the observer.*

1929 Hubble (USA) — *discovery of the relation between the distance of a galaxy and the speed at which it recedes.*

1930 Eddington (England) — *hypothesis of the expansion of the Universe.*

The increase in the distance between the galaxies can be approximately demonstrated by an inflating balloon on whose surface are randomly located several points — galaxies. The balloon increases in size, the points grow farther apart, but none occupies a privileged position.

G — our Galaxy
A, B, C, D — selected galaxies in the constellation of the Virgin (Virgo — A), Northern Crown (Corona Borealis — B), the Herdsman (Bootes — C) and the Water Snake (Hydra — D).

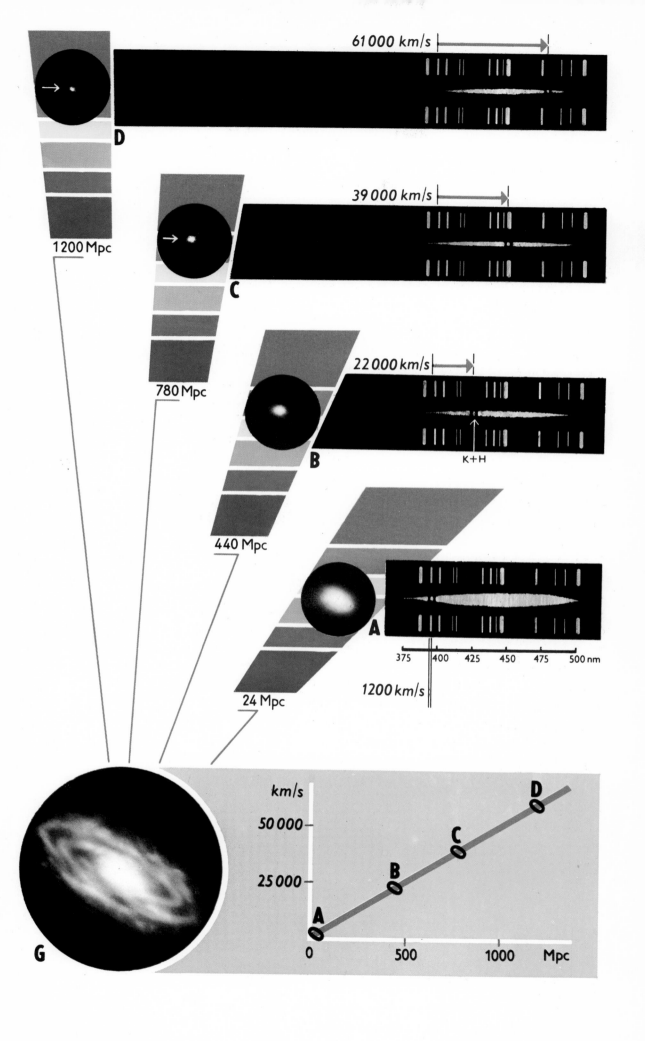

61 000 km/s

D

1200 Mpc

39 000 km/s

C

780 Mpc

22 000 km/s

B

K+H

440 Mpc

A

375    400    425    450    475    500 nm

1200 km/s

24 Mpc

G

km/s

50 000

25 000

D

C

B

A

0        500        1000        Mpc

To find one's way about amongst the stars it helps to fix in one's memory several distinctive constellations and bright stars that will then be readily recognized at any time. Using these as starting points, we can then, with the aid of star maps, venture in search of further objects in the sky. Because the appearance of the sky is forever changing owing to the motion of the Earth and its relation to the position of the observer on the Earth, we need such 'starting points' in all parts of the sky.

In the northern sky Ursa Major (UMA — the Great Bear) is our starting point for locating Polaris (the Pole Star), Leo (the Lion), and others in the direction of the lines connecting the respective stars of the 'plough' in Ursa Major. A good guide for observation of the summer sky is the Summer Triangle formed by the brightest stars in Lyra (LYR — the Lyre), Cygnus (CYG — the Swan) and Aquila (AQL — the Eagle). In the autumn sky we begin with the enlarged version of the 'plough' composed of the constellations Pegasus, Andromeda and Perseus. The winter sky is ruled over by Orion, the brightest star Sirius, and so on.

In the southern sky a good guide is the line connecting the three brightest stars: Canopus — Achernar — Fomalhaut (see also maps on pp. 155—157). The band of the Milky Way is also a help in finding our way about the heavens, and furthermore this contains the lovely constellations of Scorpius (SCO — the Scorpion), Sagittarius (SGR — the Archer), Centaurus (CEN — the Centaur) and Crux (CRU — the Southern Cross).

throughout the year constellations and groups of bright stars for purposes of general orientation

top — northern sky
bottom — southern sky
1—8 individual star maps on pp. 143—157.

*2nd century B. C. Hipparchos (Greece) compiled a catalogue of 850 stars.*
*1437—1449 Ulugh Beigh (Samarkand) — catalogue of 1,018 stars.*
*1603 Bayer (Germany) — Uranometria, star atlas which introduced the use of Greek letters for the designation of bright stars.*

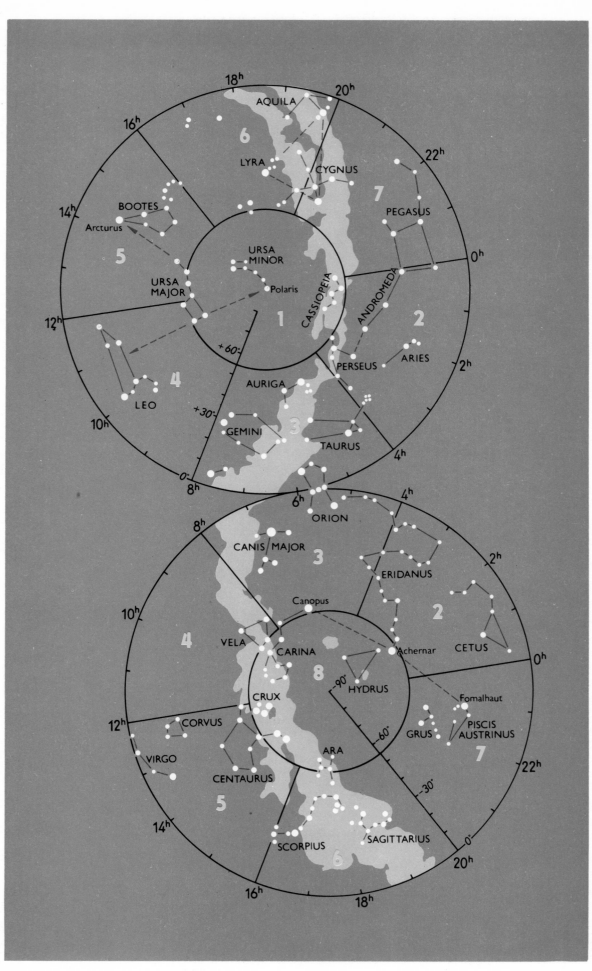

Orientation making use of the constellation of Ursa Major (the Great Bear), Polaris (the Pole Star) and the constellation of Cassiopeia. Circumpolar constellations visible above the horizon throughout the year.

Stars: α *Ursae Minoris — Polaris:* $2.0^m$, distance 466 light years, supergiant with diameter of 100 Suns, luminosity 2,000 Suns

Variables: δ *Cephei:* $3.6-4.3^m$, period 5.37 days, distance 930 light years

μ *Cephei — Erakis:* $3.6-5.1^m$, red

Double stars: α *Ursae Majoris — Dubhe:* $2.0+4.8^m$; components 0.6″ apart

ζ *Ursae Majoris — Mizar:* $2.4+4.0^m$; 14.5″, distance 78 light years

*80 Ursae Majoris — Alcor:* $4.0^m$, together with Mizar forms a triple star whose components are spectroscopic doubles; the whole is a sextuple system

Open star clusters:

χ, *h Persei:* double cluster in Perseus, visible with the unaided eye. See p. 127

Galaxies: *M 81:* spiral galaxy in UMa, apparent dimensions 16′ × 10′, distance 8.5 million light years

*M 82:* irregular galaxy in UMa, 7′ × 2′, distance 8.5 million light years, fainter than M 81

**Key**

| | |
|---|---|
| 0 1 2 3 4 5 | Apparent stellar magnitude |
| | Variable stars |
| | Double stars (Binaries) |
| | Open star clusters |
| | Globular star clusters |
| | Planetary nebulae |
| | Diffuse nebulae |
| | Galaxies |
| – – – – – | Ecliptic |

*1650 Riccioli (Italy) — discovery of the double star Mizar in the constellation Ursa Major.*

*1857 G. P. Bond (USA) — first photographed Mizar; beginnings of measuring double stars on negatives.*

*1887—1889 Pickering (USA) — Mizar resolved as a spectroscopic binary.*

**1**

12<sup>h</sup>
50°

10<sup>h</sup>

14<sup>h</sup>

ξ
Mizar Alcor
80

β

URSA    MAJOR
60°

α
Dubhe

8<sup>h</sup>

Thuban
α

70°

M81

16<sup>h</sup>

CC M82

β

LYNX

80°

URSA

DRACO

MINOR

6<sup>h</sup>

VZ

90°

ν

γ    18<sup>h</sup>

α
Polaris

CAMELOPARDALIS

80°

CEPHEUS

4<sup>h</sup>

70°

20<sup>h</sup>

μ
Erakis

α

χ    h

CASSIOPEIA

δ

Deneb

PERSEUS

60°

γ

β    ϱ

CYGNUS

2<sup>h</sup>    φ

α

22<sup>h</sup>

LACERTA

50°

0<sup>h</sup>

Winter and autumn constellations. Orientation using Pegasus (see also p. 141), Andromeda and Perseus. In the southern sky the elongate constellation of Eridanus. Located in the constellation of Pisces (the Fishes) is the vernal equinox — starting point in the system of equatorial coordinates.

Stars: α *Eridani* — *Achernar:* $0.6^m$, white, distance 78 light years

Variables: β *Persei* — *Algol:* $2.2-3.5^m$, period 68.8 hours (see p. 122)

o *Ceti* — *Mira* (Wonderful Star): $2.0-10.1^m$, period 332 days, red giant, diameter 390 Suns, distance 820 light years

Double stars: γ *Andromedae* — *Alamak:* $2.3+5.1^m$; 10″; distance 163 light years

γ *Arietis* — *Mesarthim:* $4.8+4.8^m$; 8.2″; distance 160 light years

α *Piscium* — *Alrisha:* $4.3+5.3^m$; 2″; distance 130 light years

Open star clusters:

*Pleiades:* distance 410 light years, diameter 30 light years, contains 130 stars (see p. 124)

*M 34* in Perseus: 1,450 light years, apparent diameter 35′

Planetary nebulae:

*M 76* in Perseus: 8,200 light years, diameter $157″ \times 87″$

Galaxies: *M 31* in Andromeda: 2.2 million light years (see p. 136)

*M 33* in Triangulum: 2.3 million light years, faint, nebulous formation, use small magnification for observation

**Key**

0
1
2
3
4
5

Apparent stellar magnitude

Variable stars

Double stars (Binaries)

Open star clusters

Globular star clusters

Planetary nebulae

Diffuse nebulae

Galaxies

Ecliptic

*1596 Fabricius (Holland) — determined the variability of the star omicron Ceti-Mira.*

*1784 Messier (France) — catalogue of 103 'nebulous objects', star clusters and nebulae (designation of the objects with the letter M and catalogue number, e.g. M 31, M 33, etc.).*

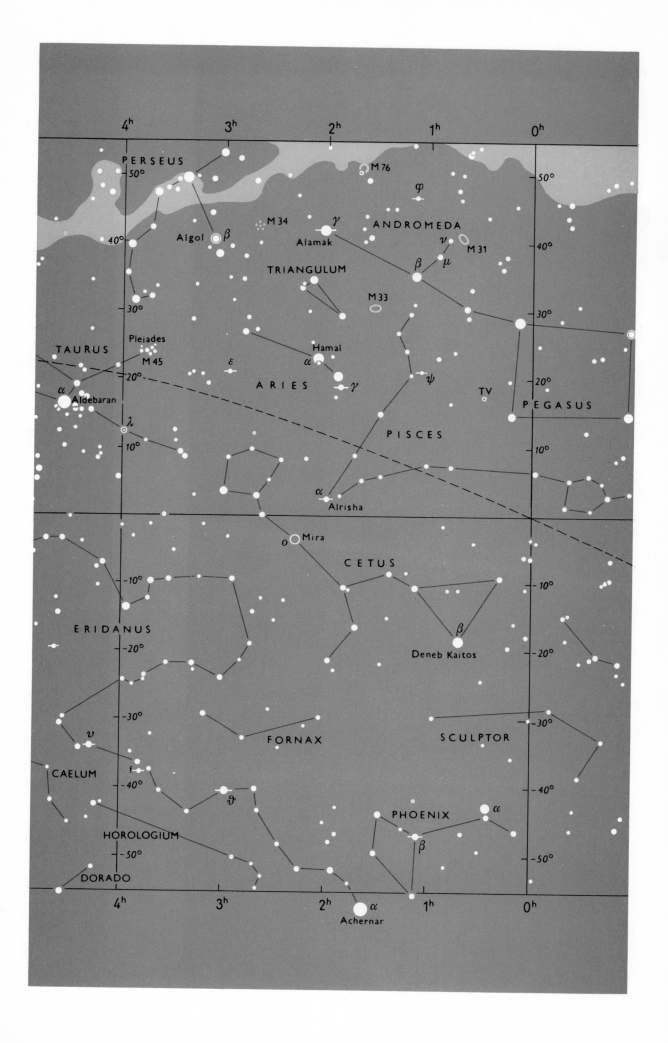

**2**

Winter constellations. Orientation from Orion and the 'winter hexagon' formed by the stars Capella, Aldebaran, Rigel, Sirius, Procyon, and Castor.

Stars: α *Aurigae — Capella:* 0.2$^m$, yellow, distance 45 light years, diameter 15 Suns
α *Tauri — Aldebaran:* 0.9$^m$, orange, distance 64 light years, diameter 36 Suns
β *Orionis — Rigel:* 0.2$^m$, white, distance 1,300 light years, diameter 48 Suns, luminosity 144,000 Suns
α *Canis Majoris — Sirius:* 1.6$^m$, white, distance 9 light years, diameter 2 Suns

Variables: α *Orionis — Betelgeuse:* 0.4—1.3$^m$, red, distance 470 light years, diameter 300 to 400 Suns
ε *Aurigae:* 3.7—4.5$^m$, period 9,883 days, distance 3,300 light years, eclipsing variable, diameter of components 300 Suns and 3,000 Suns (with atmosphere 6,000 times that of the Sun) — largest known star

Double Stars: α *Geminorum — Castor:* 2.7+2.8$^m$; 2.3″, sextuple system
β *Monocerotis:* triple star, 4.7+5.2+5.6$^m$; 7.4″ and 2.8″

Open star clusters:
*M 35* in Gemini (the Twins): distance 2,850 light years, diameter 30′, 120 stars, 9—16$^m$
*M 41* in Canis Major (the Big Dog): distance 2,500 light years, diameter 30′, 50 stars

Planetary nebulae:
*M 1* — Crab Nebula in Taurus (the Bull): distance 4,200 light years, 6′×4′, remnant of a supernova from 1054. Faint!

Diffuse nebulae:
*M 42* in Orion: very bright (see p. 132)

*1844 Bessel (Germany) proved the existence of the (then) invisible companions of the stars Sirius and Procyon.*
*1862 Clark (USA) discovered Sirius' companion.*
*1915 Adams (USA) discovered that Sirius' companion is a white dwarf — a very small and extremely dense star.*

**Key**

0
1
2
3
4
5

Apparent stellar magnitude

Variable stars

Double stars (Binaries)

Open star clusters

Globular star clusters

Planetary nebulae

Diffuse nebulae

Galaxies

Ecliptic

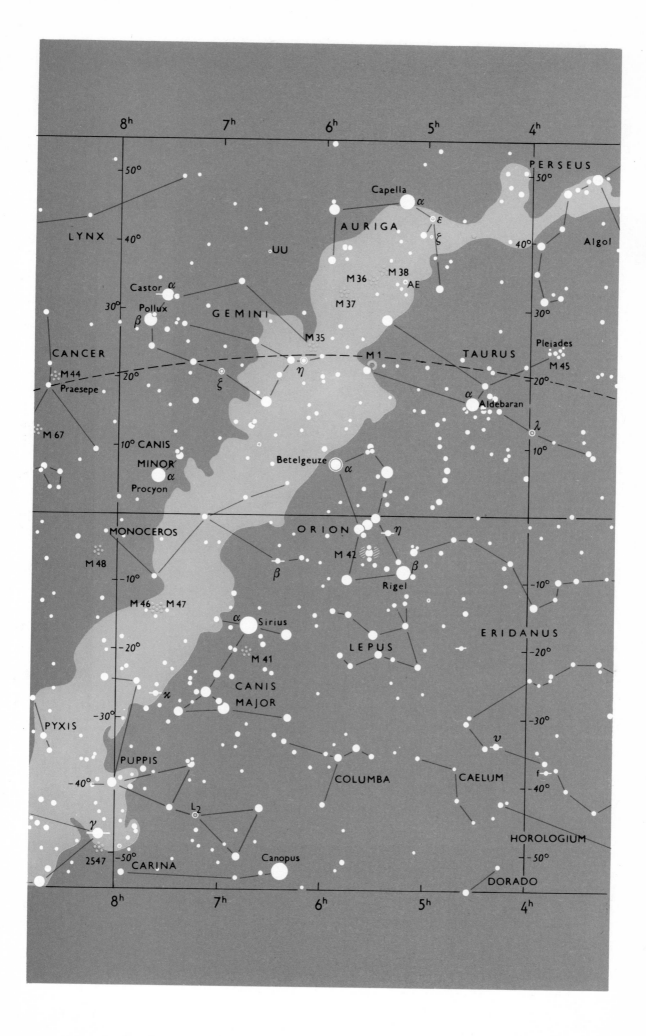

# RIGHT ASCENSION
$8^h - 12^h$

Spring constellations. Orientation around Leo (the Lion), lying south of Ursa Major (the Great Bear, see also p. 143). South of the constellation of Cancer (the Crab) is the head of Hydra (the Water Snake).

Stars: α *Leonis* — *Regulus:* 1.3$^m$, white, distance 68 light years

β *Leonis* — *Denebola:* 2.2$^m$, white, distance 42 light years

Variables: *R Leonis:* 5.4—10.5$^m$, period 313 days, Mira Ceti type

Double stars: γ *Leonis* — *Algieba:* 2.6 + 3.8$^m$; 4.3″; orange and yellow, distance 130 light years

γ *Velorum:* 4.8 + 2.2$^m$; 41″; both components are spectroscopic doubles, distance 650 light years

Open star clusters:

*M 44* — Praesepe (Beehive): distance 525 light years, diameter 1.5°, 60 stars, age 400 million years; see p. 126

*M 67* in Cancer (the Crab): distance 2,700 light years, diameter 27′ (14 light years), 500 stars 9—15$^m$, age 4 to 5 thousand million years, oldest of the open star clusters

Galaxies: *M 65, M 66, M 95, M 96* — spiral galaxies in Leo (the Lion), members of a cluster of galaxies, distant approximately 29 million light years. Brightness of galaxies from 8.4$^m$ to 10.4$^m$, diameters 5′ to 8′, true diameters 80 to 120,000 light years

This part of the sky contains 'galactic windows' where interstellar matter does not hinder our view of the outermost regions of space.

Key

0 — 

1 — 

2 — Apparent stellar magnitude

3 — 

4 — 

5 — 

Variable stars

Double stars (Binaries)

Open star clusters

Globular star clusters

Planetary nebulae

Diffuse nebulae

Galaxies

Ecliptic

*1912—1914 Slipher (USA) — first measured the speed at which galaxies are moving away from us.*

*1925 Hubble (USA) introduced the classification of galaxies as elliptical, spiral and irregular.*

*1963 M. Schmidt (USA) — discovery of the first quasar.*

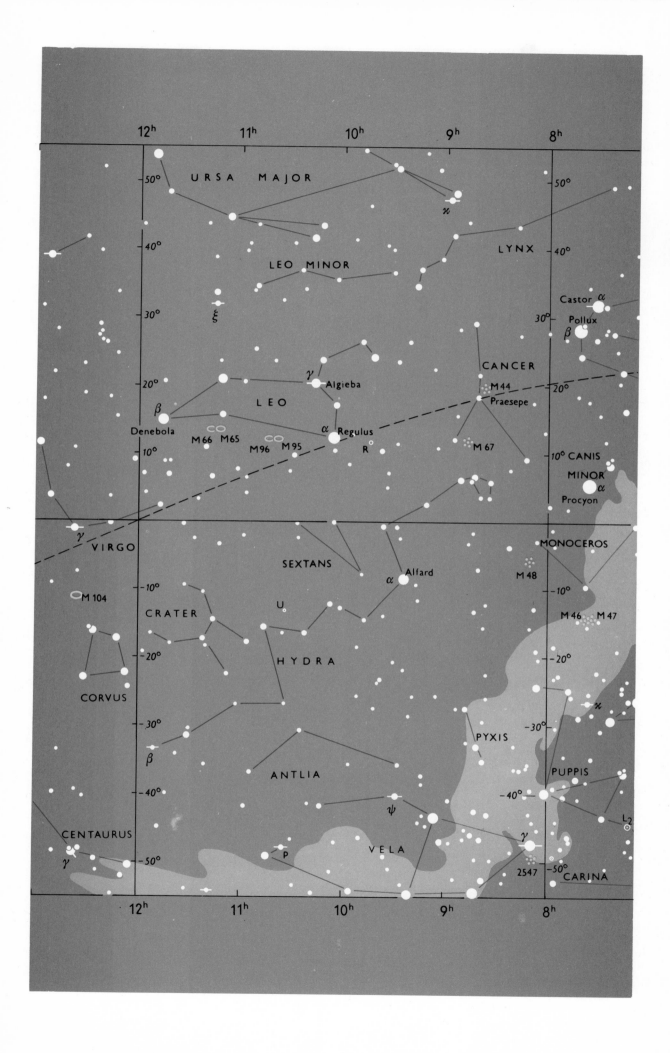

Spring constellations. Orientation using the Spring Triangle formed by the three stars Regulus (see p. 149), Arcturus and Spica.

Stars: α *Bootis* — *Arcturus:* $0.0^m$, orange, distance 35 light years, diameter 23 Suns, luminosity 180 Suns
α *Virginis* — *Spica:* $1.0^m$, white, distance 218 light years

Variables: *R Hydrae:* $4.0 - 10.0^m$, period 386 days, Mira Ceti type

Double stars: α *Canum Venaticorum* — *Cor Caroli (Charles' Heart):*
$2.9 + 5.4^m$, 19.7″, distance 130 light years (see p. 121)
ε *Bootis* — *Izar:* $2.7 + 5.1^m$; 2.9″; distance 170 light years (see p. 121)
γ *Virginis* — *Porrima:* $3.6 + 3.6^m$; 5.2″; distance 37 light years (see p. 121)

Globular clusters:
*M 3* in Canes Venatici (the Hunting Dogs): distance 48,500 light years, diameter 10′ (325 light years)
*M 5* in Serpens (the Serpent): distance 27,000 light years, diameter 15′ (130 light years)
ω *Centauri:* distance 16,000 light years, diameter 23′ (150 light years)

Galaxies: *M 51* in Canes Venatici: $8.1^m$, distance 37 million light years, diameter $12′ \times 6′$ (125,000 light years) — whirlpool galaxy
*M 101* in Ursa Major (the Great Bear): distance 11.5 million light years, diameter 22′ (92,000 light years), spiral galaxy
*M 104* — Sombrero in Virgo (the Virgin): $8.7^m$, distance 41 million light years, diameter $6′ \times 3′$ (142,000 light years)

**Key**

| | |
|---|---|
| 0 | |
| 1 | Apparent stellar magnitude |
| 2 | |
| 3 | |
| 4 | |
| 5 | |

Variable stars

Double stars (Binaries)

Open star clusters

Globular star clusters

Planetary nebulae

Diffuse nebulae

Galaxies

- - - - - Ecliptic

*1859—1862 Argelander (Germany)* — Bonner Durchmusterung *catalogue of 324,000 stars.*
*1888 Dreyer (USA)* — New General Catalogue (NGC) *containing 7,840 nebulae.*

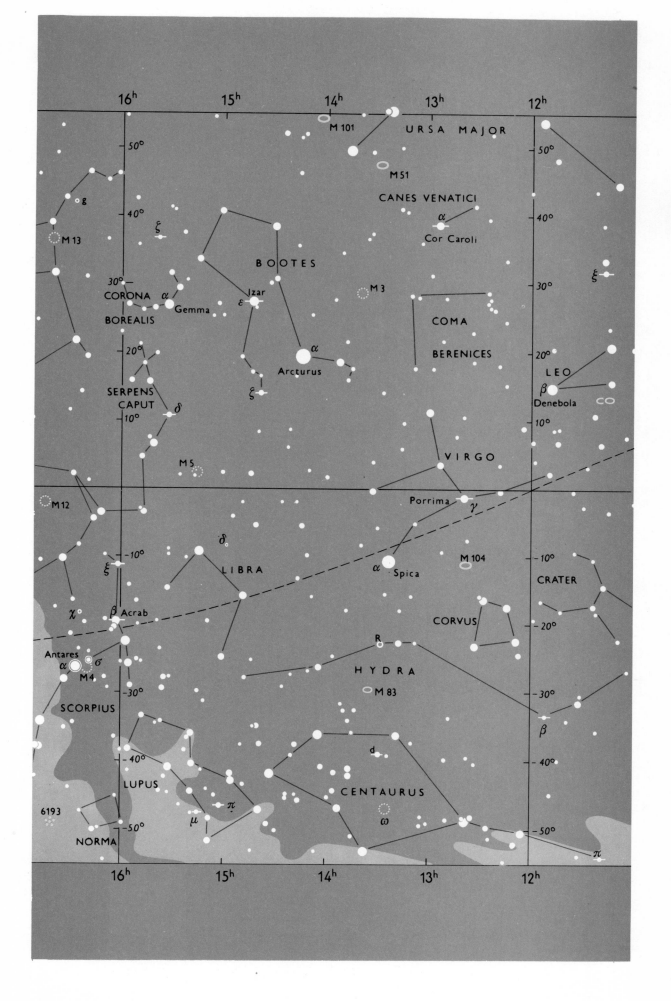

5

Summer constellations. The brightest parts of the Milky Way (see also p. 113). Orientation using the Summer Triangle formed by the stars Vega, Deneb (p. 154) and Atair.

Stars: α *Lyrae* — *Vega:* $0.0^m$, white, distance 26 light years, diameter 3.2 Suns

    α *Aquillae* — *Atair:* $0.9^m$, white, distance 16 l. y.

    α *Scorpii* — *Antares:* $1.0^m$, red, distance 375 light years, supergiant, diameter 420 Suns, luminosity 1,900 Suns

Variables: β *Lyrae:* $3.4 - 4.3^m$, period 12 days 22.4 hours; distance 1,100 light years, close eclipsing binary

    η *Aquilae:* $3.7 - 4.4^m$, period 7.2 days, cepheid

Double stars: β *Cygni* — *Albireo:* $3.2 + 5.4^m$; 34.6″; see p. 121

    β *Scorpii* — *Acrab:* $2.9 + 5.1^m$; 14″; white and greenish-yellow

    α *Herculis* — *Ras Algethi:* 3.0 to $4.0^m + 5.4^m$; 4.6″; see p. 121 (the brighter component is a variable)

    ε *Lyrae:* quadruple star — see p. 121

Open star clusters:

    *M 11* in Scutum (the Shield): $6.3^m$, distance 5,500 light years, diameter 20′ (18 light years)

Globular star clusters:

    *M 13* in Hercules: see p. 127

    *M 4* in Scorpius (the Scorpion): $6.4^m$, distance 7,500 light years, diameter 14′ (95 light years)

Planetary nebulae:

    *M 57* in Lyra (the Lyre): Ring Nebula — see p. 128

    *M 27* in Vulpecula (the Fox): Dumb-bell — p. 128

Diffuse nebulae:

    *M 8* — Lagoon, *M 17* — Omega, *M 20* — Trifid: faint nebulae in Sagittarius that are difficult objects for visual observation.

*1850 Whipple (USA) photographed Vega — beginnings of astrophotography.*

*1949 Kalinyak, Krasovskii, Nikonov (USSR) — first infrared photographs of the nucleus of the Galaxy: the nucleus has a diameter of approximately 4,000 light years.*

**Key**

| | |
|---|---|
| 0 | |
| 1 | Apparent stellar |
| 2 | magnitude |
| 3 | |
| 4 | |
| 5 | |

Variable stars

Double stars (Binaries)

Open star clusters

Globular star clusters

Planetary nebulae

Diffuse nebulae

Galaxies

Ecliptic

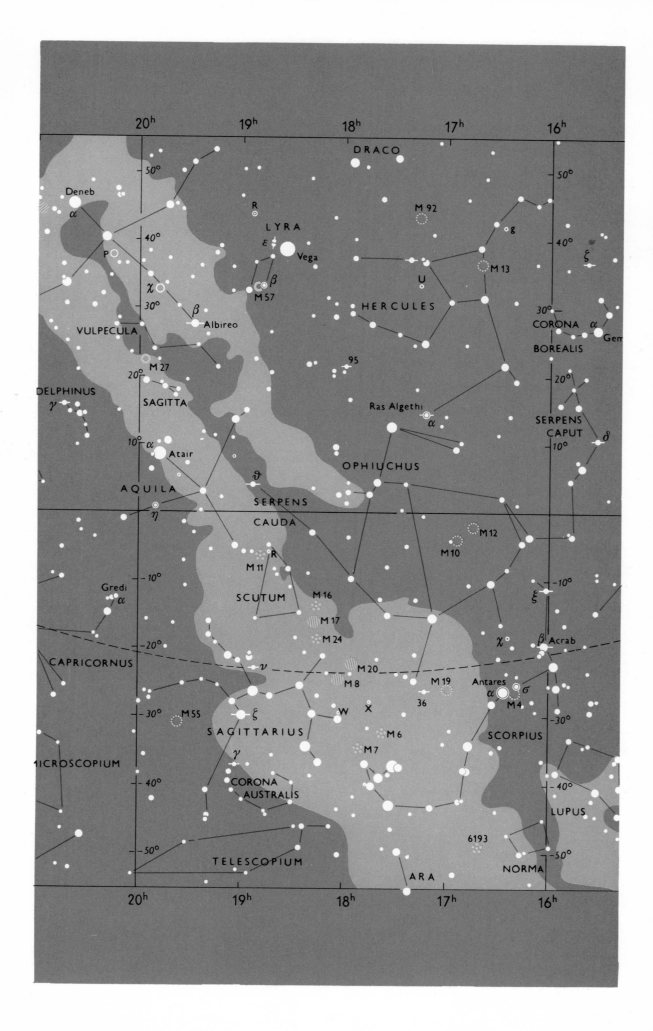

20ʰ  19ʰ  18ʰ  17ʰ  16ʰ

DRACO

50°

Deneb
α
40°
P
χ
30°
β
VULPECULA
Albireo

R
LYRA
ε
Vega
β
M 57

M 92
g
U
M 13
HERCULES

50°
ξ
40°
30°
CORONA α
BOREALIS Gem

M 27
20°
SAGITTA

DELPHINUS
γ

10° α
Atair
AQUILA
η

95

Ras Algethi
α

OPHIUCHUS

20°
SERPENS
CAPUT
δ
10°

SERPENS
CAUDA
R
M 11
-10°
SCUTUM
M 16
M 17
M 24

Gredi
α

M 12
M 10

ξ
-10°

-20°
CAPRICORNUS
ν
M 20
M 8

W  X

36

χ
β Acrab

Antares
α  σ
M 4
-30°

M 55
-30° ξ
SAGITTARIUS
γ

M 19

M 6
M 7

SCORPIUS

-40°
LUPUS

MICROSCOPIUM

CORONA
AUSTRALIS

-40°

-50°
TELESCOPIUM

ARA

6193
-50°

NORMA

20ʰ  19ʰ  18ʰ  17ʰ  16ʰ

6

Summer and autumn constellations. Orientation according to the 'Great Square of Pegasus' marked by the stars α, β and γ in Pegasus and α in Andromeda. In the southern sky the bright star Fomalhaut in Piscis Austrinus (the Southern Fish).

Stars: α *Cygni — Deneb:* 1.3$^m$, white, distance 930 light years, luminosity 20,000 Suns

α *Piscis Austrini — Fomalhaut:* 1.2$^m$, white, distance 23 light years, luminosity 10 Suns

Variables: β *Pegasi — Scheat:* 2.4—2.8$^m$, irregular variable, distance 170 light years, diameter 110 Suns

Double stars: γ *Delphini:* 4.5+5.5$^m$; 10.4″; p. 121

ζ *Aquarii:* 4.4+4.6$^m$; 2″

α$_1$ α$_2$ *Capricorni — Gredi:* optical binary which can be resolved with the eye; α$_1$: 3.8$^m$, distance 1,100 light years; α$_2$: 4,5$^m$, distance 116 light years
angular distance of α$_1$α$_2$: 6.4′

Open star clusters:

*M 39* in Cygnus (the Swan): 5.2$^m$, distance 815 light years, diameter 30′; 25 stars

Globular star clusters:

*M 2* in Aquarius (the Water Carrier): 6.3$^m$, distance 55,000 light years, diameter 12′ (105 light years)

*M 15* in Pegasus: 6.0$^m$, distance 49,000 light years, diameter 12′ (90 light years)

Planetary nebulae:

*NGC 7293* — Helix in Aquarius: see p. 128

Diffuse nebulae:

*NGC 7000* — America in Cygnus: distance 800 light years, visible with unaided eye or binoculars, if conditions are excellent, as a small hazy cloud about 2° in diameter east of Deneb: see p. 113

| Key | |
|---|---|
| 0 ● | |
| 1 ● | Apparent stellar |
| 2 ● | magnitude |
| 3 ● | |
| 4 • | |
| 5 · | |
| ◎ ○ | Variable stars |
| —•— | Double stars (Binaries) |
| ∷ | Open star clusters |
| ○ | Globular star clusters |
| ◉ | Planetary nebulae |
| ▨ | Diffuse nebulae |
| ⬭ | Galaxies |
| – – – – | Ecliptic |

*1961 Discovery of cosmic gamma radiation.*
*1963 Discovery of the first source of X-ray radiation (in the Crab Nebula).*
*1968—1969 J. Bell and A. Hewish (England) — discovery of pulsars.*

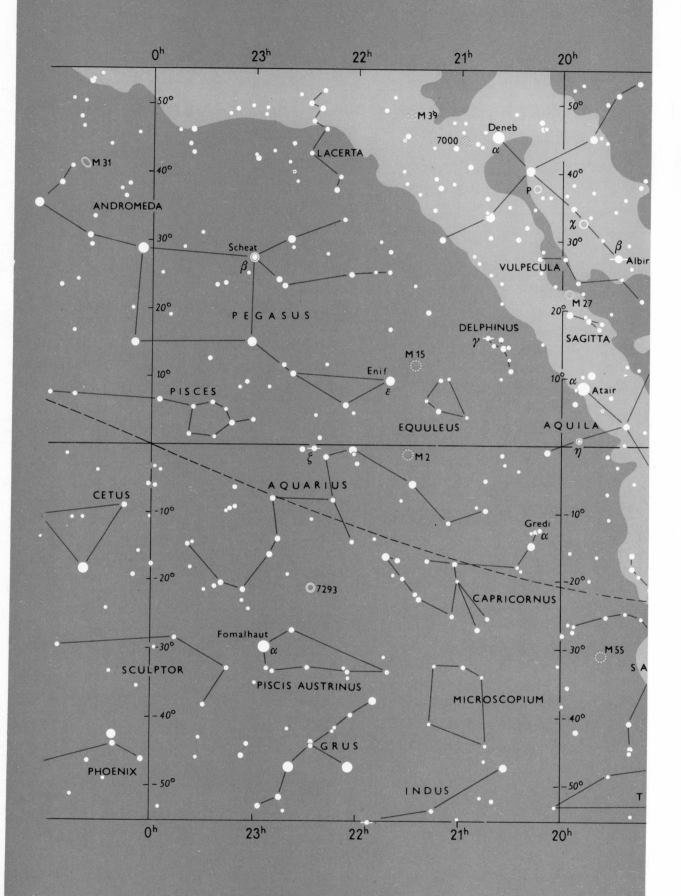

# REGION
# OF THE SOUTH POLE

Orientation according to the constellation of Crux (the Southern Cross); its longer arm points to the south celestial pole. The line connecting α and β Centauri points to this cross, thus helping to distinguish it from the false cross on the boundaries of Carina and Vela.

Stars: α *Carinae* — *Canopus:* $0.9^m$, white, distance 365 light years
   α *Eridani* — *Achernar:* $0.6^m$, white, distance 78 light years
Double stars: α *Centauri* — *Toliman:* triple star $0.3^m + 1.7^m$; 35″; faintest component $11.3^m$ (Proxima Centauri), distance 4.3 light years
   β *Tucanae:* $4.5 + 4.5^m$; 27.1″
Open star clusters:
   *NGC 3532* in Carina: $3.3^m$, distance 1,700 light years diameter 60′
   *NGC 4755* close to β Crucis: $5.2^m$, distance 980 light years, diameter 10′
Globular star clusters:
   *47 Tucanae* (= NGC 104): distance 19,000 light years, diameter 23′
   *NGC 6752* in Pavo (the Peacock): distance 20,000 light years, diameter 13′
   *NGC 6397* in Ara (the Altar): distance 7,500 light years, diameter 19′
Nebulae: 'The Coalsack' in the Southern Cross — dark dust cloud 500 light years distant
   *NGC 2070* — Tarantula — diffuse nebula in the LMC, angular dimensions 20′ × 20′
Galaxies: *LMC* in Dorado (Large Magellanic Cloud): distance 200,000 light years, diameter 9°
   *SMC* in Tucana (Small Magellanic Cloud): distance 220,000 light years, diameter 5°

*1912 Leawitt (USA) — discovered a relation between the period and luminosity of 25 cepheids in SMC.*
*1913 Hertzsprung (Denmark) — determined the distance of SMC from the period-luminosity relation of cepheids.*

**Key**

| | |
|---|---|
| 0 ● | |
| 1 ● | |
| 2 ● | Apparent stellar magnitude |
| 3 • | |
| 4 · | |
| 5 · | |
| ◎ ○ | Variable stars |
| ● | Double stars (Binaries) |
| ⁙ | Open star clusters |
| ⦂ | Globular star clusters |
| ◉ | Planetary nebulae |
| ▨ | Diffuse nebulae |
| ⬭ | Galaxies |
| − − − − − | Ecliptic |

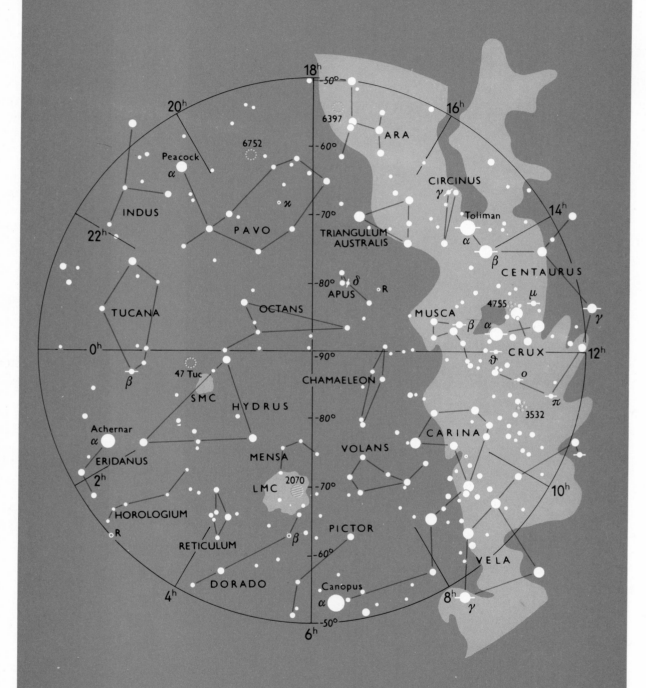

# THE MOON AND PLANETS
## IN THE YEARS 1979 — 2000

For his observations the amateur astronomer often needs to have on hand ready information pertaining to the visibility of the Moon and planets during a certain period or on a certain night. Though far from being a substitute for an astronomical year-book, which contains all necessary data, the following few pages of graphic diagrams providing the reader with a great deal of useful information should prove a handy aid.

From these diagrams it is possible to determine the conditions for observing the five brightest planets with the unaided eye or telescope, their angular distance from the Sun and whether they can be observed in the evening, morning or throughout the night, when they are at conjunction or opposition to the Sun, when Mercury and Venus are at their greatest elongation, the constellation in which the respective planet is located, and so on. From the tables giving the Moon's phases it is easy to determine the age of the Moon at any given date and with the aid of the diagrams to estimate the position of the Moon in the heavens. The tables giving the Moon's phases also contain data on all eclipses of the Sun and Moon visible from the Earth in the years 1979—2000.

**Description of the Diagrams**
The vertical axis of the diagram marks the position of the Sun from which the elongation (angular distance) of the planet is measured to the east (E) or west (W), always from 0° to 180°. Every 30° of elongation is marked by a long vertical line and every 10° by a short vertical mark. If, in such a diagram, a planet is to the left (east) of the vertical axis (the Sun), it means that it sets after the Sun and is therefore visible in the evening sky. Conversely, a planet located in the diagram to the right of the Sun, i. e. at western elongation, rises before the Sun and can be observed in the morning sky. If a planet is at the left or right edge of the diagram, 180° from the Sun, then it is at opposition to the Sun and can be observed the whole night.

The vertical yellow band marks the zone 10° east and west of the Sun where, as a rule, the planets cannot be observed.

The diagonal coloured bands denote the constellations along the ecliptic; the edges of these bands mark the boundaries between the constellations (boundaries accepted by the IAU and printed on standard detailed stellar maps). We call attention to the fact that represented in the diagrams are the *constellations* and not the so-called signs of the zodiac (regular, 30° portions of the ecliptic) and that therefore some bands are broader and others narrower, according to the portion of the ecliptic occupied by the respective constellation. The constellations of the Scorpion (Scorpio — SCO) and Serpent-bearer (Ophiuchus — OPH) are designated by a common band; although the Serpent-bearer is not included among the zodiacal constellations, it occupies a portion of the ecliptic that is three times as long as that occupied by the Scorpion.

Constellations of the northern sky, which rise higher above the horizon for the northern observer and may be observed the greater part of the night, are marked in red; constellations in the southern part of the ecliptic, which rise lower above the northern observer's horizon and can be observed for a shorter time, are marked in blue.

The meaning of the international abbreviations for the names of the constellations is as follows:

| | |
|---|---|
| ARI | Aries — Ram |
| TAU | Taurus — Bull |
| GEM | Gemini — Twins |
| CNC | Cancer — Crab |
| LEO | Leo — Lion |
| VIR | Virgo — Virgin |
| LIB | Libra — Scales |
| SCO | Scorpius — Scorpion |
| OPH | Ophiuchus — Serpent-bearer |
| SGR | Sagittarius — Archer |
| CAP | Capricornus — Goat |

AQR Aquarius — Water-Carrier
PSC Pisces — Fishes

The five brightest planets are designated in the diagrams by different kinds of lines and the following symbols:

H — Mercury (Hermes)
V — Venus
M — Mars
J — Jupiter
S — Saturn

At the point where the lines of the planets intersect the respective planets are in conjunction. The conjunctions in each year are numbered in chronological order (except for conjunctions within 10° of the Sun) and also arranged in a special table to the right of the diagram which gives more detailed information: the date and hour (U. T.) of conjunction at right ascension and the mutual angular distance of the two planets in degrees; the planet listed first is at conjunction north of the other planet.

The tables giving the phases of the Moon, on the pages opposite the diagrams, make it easy to estimate the phase of the Moon and its position in the sky for any given date. It should be kept in mind that in the first quarter the Moon is always at eastern elongation 90° from the Sun, at full Moon it is 180° from the Sun (both in the sky and in the diagram), in the last quarter it is at western elongation 90° from the Sun, and the new Moon is at an elongation of 0°. The elongation of the Moon increases an average of 12.2° every day. On the basis of these data it is possible to determine the approximate elongation of the Moon on a specific date and estimate its position from the diagram accordingly.

The tables giving the phases of the Moon likewise make it possible approximately to determine the age of the Moon on any given date in the years 1979—2000 and by referring to the text and map on p. 44 to estimate roughly the position of the terminator, illumination of the Moon's surface, etc.

The following are two examples for the year 1979 which help explain how to use the diagrams.

Example 1

**What are the general conditions for observation of the planets during the course of the year 1979?**

*Mercury* may be observed as the morning star at the beginning of January, throughout April and the beginning of May, in mid-August and in the first half of December. It may be observed as the evening star in the first half of March, from mid-June to mid-July and from mid-October until the beginning of November.

The best time for locating Mercury is when it is at its greatest eastern elongation in spring and greatest western elongation in autumn. Another good time for locating Mercury in the sky is when it is at conjunction with a brighter planet, e. g. with Jupiter on 30 August or Venus on 8 November 1979. Conditions are also good at the beginning of July, when for many days Mercury is only several degrees from Jupiter.

*Venus* is in the morning sky from the beginning of the year, in mid-January it is at its greatest western elongation. It may be observed as the morning star until mid-July, when it disappears in the light of the rising Sun. At the end of August it is at superior conjunction with the Sun and at the beginning of October it appears in the evening sky, where it remains until the end of the year. Comparison with the illustration on p. 73 gives us an idea of what the phases of Venus and Mercury look like during the course of the year. Venus is most brilliant 35 days before and 35 days after inferior conjunction; in the following diagram we find that the earliest inferior conjunction will be in June 1980.

*Mars* cannot be observed at the beginning of 1979 because it is in conjunction with the Sun in January. At the end of March it appears in the morning sky, where it remains until the end of the year. As its western elongation increases, so does the length of time Mars can be observed above the horizon —

The inferior and superior conjunction of Mercury or Venus with the Sun can be distinguished in the diagrams according to the direction in which the inferior planet is moving. The planet passes through superior conjunction from the morning to the evening sky (from right to left in the diagram) and through inferior conjunction in the opposite direction.

at the end of the year it rises before midnight. Mars' apparent motion amidst the stars is rapid; on 20 May it is in conjunction with Venus in the constellation of Aries, although it is 1° north of Venus. In June and July Mars is in Taurus, in August and September in Gemini, up to mid-October in Cancer and until the end of the year in Leo, where it encounters Jupiter.

*Jupiter* is at opposition with the Sun on 24 January and therefore above the horizon throughout the night at the beginning of the year; it moves slowly amidst the stars on the boundary between Gemini and Cancer and rises high above the horizon together with these constellations. At the end of April Jupiter is already less than 90° east of the Sun and therefore sets around midnight. In early summer it is in Cancer and visible only for a brief period after sunset; in August it cannot be observed because on 13 August it is in conjunction with the Sun. At the beginning of September Jupiter appears in the morning sky in the constellation of Leo, where it remains until the end of the year. Jupiter is in conjunction with Mars on 13 December at 17 h U. T.; at this time Mars passes 1.7° north of Jupiter.

*Saturn* is in the constellation of Leo throughout the year and at opposition with the Sun on 1 March. Conditions for observing Saturn are similar to those for observing Jupiter because the two planets are close to each other and come increasingly closer during the course of the year; this approximation will continue until the year 1981 (see diagrams on p. 163), when Jupiter will encounter Saturn three times. This will be the last triple conjunction of the two planets in the twentieth century. Saturn's rings are practically closed in 1979 (see p. 97) and at the end of October observers will be unable to see them for a brief period, because at that time the Earth will be crossing the plane of the rings.

### Example 2
**What are the conditions for observing the Moon and the planets on 1 November 1979?**

*Mercury* and *Venus* can be observed for a brief period after sunset as evening stars. Mercury is near its greatest elongation, approximately 24° east of the Sun, and thus its phase appears somewhat like that of the Moon shortly after the last quarter. By 1 November Venus has travelled roughly the first third of its orbit from superior conjunction to eastern elongation, so that its phase will be like that of the Moon shortly after Full Moon. To the observer Mercury and Venus appear

| 1979 (months) | 1 | 2 | 3 | 4 | 5 | 6 | 7 | 8 | 9 | 10 | 11 | 12 |
|---|---|---|---|---|---|---|---|---|---|---|---|---|
| first quarter | 5 | 4 | 5 | 4 | 4 | 2 | 2 | 1 | | | | |
| Full Moon | 13 | 12 | 13 | 12 | 12 | 10 | 9 | 8 | 6 | 5 | 4 | 3 |
| last quarter | 21 | 20 | 21 | 19 | 18 | 17 | 16 | 14 | 13 | 12 | 11 | 11 |
| New Moon | 28 | 26 | 28 | 26 | 26 | 24 | 24 | 22 | 21 | 21 | 19 | 19 |
| first quarter | | | | | | | | 30 | 29 | 28 | 26 | 26 |

ECLIPSE OF THE MOON *13 March: partial* — Africa, Europe, Australia, Asia; *6 September: total* — Pacific, east Asia, western North America, Australia
ECLIPSE OF THE SUN *26 February: total* — North and Central America, Greenland; *22 August: annular* — southeast Pacific, southern South America, southwest Atlantic
Note: on 27 October Earth crosses the plane of Saturn's rings from south to north.

to be close together (approximately 7°) and they will be closest to each other on 8 November (conjunction no. 5).

The planets *Mars, Jupiter* and *Saturn* rise, in this order, above the horizon after midnight, and in the morning sky all three are in the constellation of Leo, where they are displaced at practically identical intervals of 15°. At this time Mars' disc still has a negligible angular diameter because there are practically 4 months to go until it is at opposition with the Sun. Saturn's rings are 'closed' and cannot be observed with a small telescope.

In the tables giving the phases of the Moon for the year 1979 we find that on 1 November the Moon is 3 days before Full Moon. This means that the age of the Moon is about 12 days and that its elongation is approximately 140° east of the Sun (12 × 12.2°), which is in the constellation of Pisces, as we can see from the diagram.

*Bibliography:* Jean Meeus, *Memoirs 5,* Ver. v. Sterrenkunde, Brussels, Belgium (numerical values used with the kind permission of the author)

*Explanatory notes to the tables on the conjunctions of planets:*

a  number of the conjunction in the given year
b  month and day
c  hour (U.T.) of conjunction in right ascension
d  planet passing north
e  angular distance in degrees
f  planet passing south
g  elongation in degrees: E — eastern, W — western

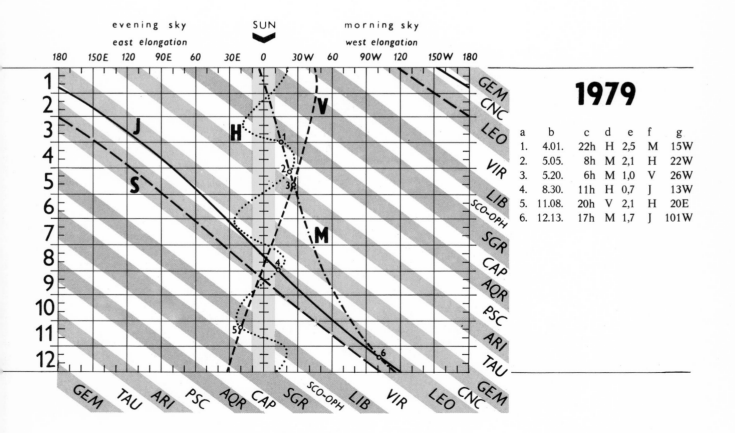

# 1979

| a | b | c | d | e | f | g |
|---|---|---|---|---|---|---|
| 1. | 4.01. | 22h | H | 2,5 | M | 15W |
| 2. | 5.05. | 8h | M | 2,1 | H | 22W |
| 3. | 5.20. | 6h | M | 1,0 | V | 26W |
| 4. | 8.30. | 11h | H | 0,7 | J | 13W |
| 5. | 11.08. | 20h | V | 2,1 | H | 20E |
| 6. | 12.13. | 17h | M | 1,7 | J | 101W |

| 1980 | 1 | 2 | 3 | 4 | 5 | 6 | 7 | 8 | 9 | 10 | 11 | 12 |
|---|---|---|---|---|---|---|---|---|---|---|---|---|
| Full Moon | 2 | 1 | 1 | | | | | | | | | |
| last quarter | 10 | 9 | 9 | 8 | 7 | 6 | 5 | 3 | 1 | 1 | | |
| New Moon | 17 | 16 | 16 | 15 | 14 | 12 | 12 | 10 | 9 | 9 | 7 | 7 |
| first quarter | 24 | 23 | 23 | 22 | 21 | 20 | 20 | 18 | 17 | 17 | 15 | 15 |
| Full Moon | | | 31 | 30 | 29 | 28 | 27 | 26 | 24 | 23 | 22 | 21 |
| last quarter | | | | | | | | | | 30 | 29 | 29 |

ECLIPSE OF THE SUN *16 February: total* — Africa, south and southeast Asia; *10 August: annular* — Central America, South America
Note: on 12 March Earth crosses the plane of Saturn's rings from north to south; a second time on 23 July from south to north; on 3 March the Sun crosses the plane of Saturn's rings from south to north.

| 1981 | 1 | 2 | 3 | 4 | 5 | 6 | 7 | 8 | 9 | 10 | 11 | 12 |
|---|---|---|---|---|---|---|---|---|---|---|---|---|
| New Moon | 6 | 4 | 6 | 4 | 4 | 2 | 1 | | | | | |
| first quarter | 13 | 11 | 13 | 11 | 10 | 9 | 9 | 7 | 6 | 6 | 5 | 4 |
| Full Moon | 20 | 18 | 20 | 19 | 19 | 17 | 17 | 15 | 14 | 13 | 11 | 11 |
| last quarter | 28 | 27 | 28 | 27 | 26 | 25 | 24 | 22 | 20 | 20 | 18 | 18 |
| New Moon | | | | | | | 31 | 29 | 28 | 27 | 26 | 26 |

ECLIPSE OF THE MOON *17 July: partial* — South America, North America, Antarctic, west Africa
ECLIPSE OF THE SUN *4 February: annular* — southeast Australia, Antarctic, South America; *31 July: total* — Asia, Arctic, northwestern North America
Note: in 1981 — triple conjunction of Jupiter and Saturn; the last was in 1940—1941, the next will be in 2238—2239

| 1982 | 1 | 2 | 3 | 4 | 5 | 6 | 7 | 8 | 9 | 10 | 11 | 12 |
|---|---|---|---|---|---|---|---|---|---|---|---|---|
| first quarter | 3 | 1 | 2 | 1 | | | | | | | | |
| Full Moon | 9 | 8 | 9 | 8 | 8 | 6 | 6 | 4 | 3 | 3 | 1 | 1 |
| last quarter | 16 | 15 | 17 | 16 | 16 | 14 | 14 | 12 | 10 | 9 | 8 | 7 |
| New Moon | 25 | 23 | 25 | 23 | 23 | 21 | 20 | 19 | 17 | 17 | 15 | 15 |
| first quarter | | | | 30 | 29 | 28 | 27 | 26 | 25 | 25 | 23 | 23 |
| Full Moon | | | | | | | | | | | | 30 |

ECLIPSE OF THE MOON *9 January: total* — Asia, Arctic, Greenland, Indian Ocean, western Australia; *6 July: total* — North and South America; *30 December: total* — east Asia, Australia, North America
ECLIPSE OF THE SUN *25 January: partial* — southern Indian Ocean, Antarctic, southwest Pacific; *21 June: partial* — south Atlantic, south Africa, southwestern Indian Ocean; *July: partial* — northeast Asia, northwestern Europe, northern North America; *15 December: partial* — Europe, north Africa, southwest Asia

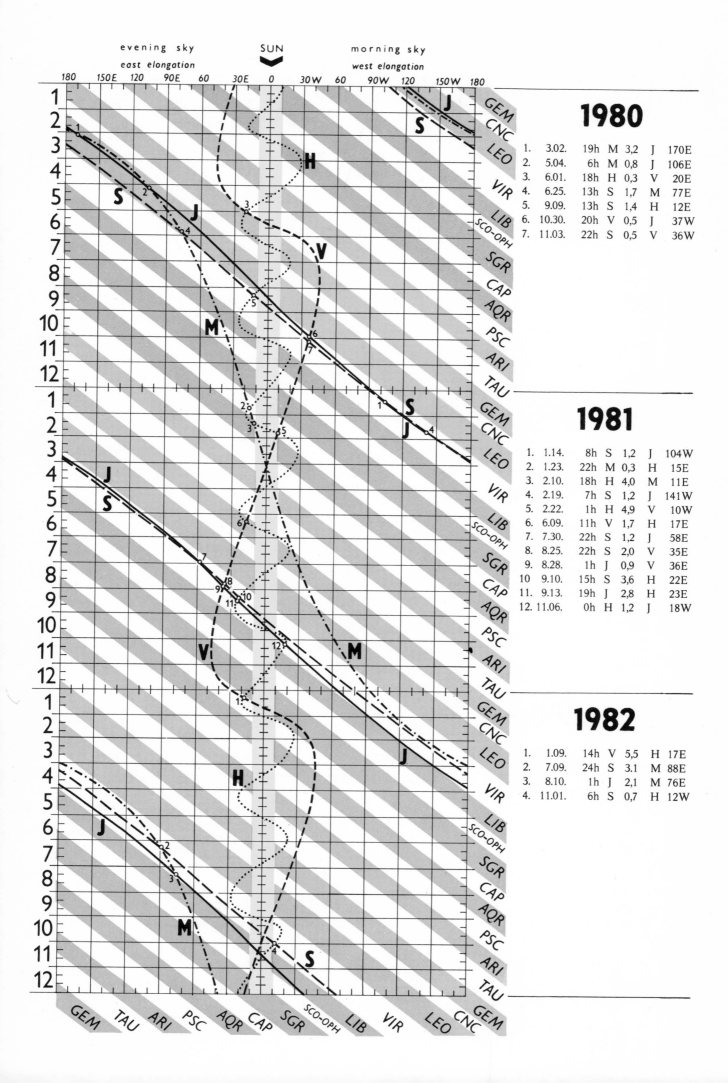

**1980**

| | | | | | | |
|---|---|---|---|---|---|---|
| 1. | 3.02. | 19h | M | 3,2 | J | 170E |
| 2. | 5.04. | 6h | M | 0,8 | J | 106E |
| 3. | 6.01. | 18h | H | 0,3 | V | 20E |
| 4. | 6.25. | 13h | S | 1,7 | M | 77E |
| 5. | 9.09. | 13h | S | 1,4 | H | 12E |
| 6. | 10.30. | 20h | V | 0,5 | J | 37W |
| 7. | 11.03. | 22h | S | 0,5 | V | 36W |

**1981**

| | | | | | | |
|---|---|---|---|---|---|---|
| 1. | 1.14. | 8h | S | 1,2 | J | 104W |
| 2. | 1.23. | 22h | M | 0,3 | H | 15E |
| 3. | 2.10. | 18h | H | 4,0 | M | 11E |
| 4. | 2.19. | 7h | S | 1,2 | J | 141W |
| 5. | 2.22. | 1h | H | 4,9 | V | 10W |
| 6. | 6.09. | 11h | V | 1,7 | H | 17E |
| 7. | 7.30. | 22h | S | 1,2 | J | 58E |
| 8. | 8.25. | 22h | S | 2,0 | V | 35E |
| 9. | 8.28. | 1h | J | 0,9 | V | 36E |
| 10 | 9.10. | 15h | S | 3,6 | H | 22E |
| 11. | 9.13. | 19h | J | 2,8 | H | 23E |
| 12. | 11.06. | 0h | H | 1,2 | J | 18W |

**1982**

| | | | | | | |
|---|---|---|---|---|---|---|
| 1. | 1.09. | 14h | V | 5,5 | H | 17E |
| 2. | 7.09. | 24h | S | 3.1 | M | 88E |
| 3. | 8.10. | 1h | J | 2,1 | M | 76E |
| 4. | 11.01. | 6h | S | 0,7 | H | 12W |

| 1983 | 1 | 2 | 3 | 4 | 5 | 6 | 7 | 8 | 9 | 10 | 11 | 12 |
|---|---|---|---|---|---|---|---|---|---|---|---|---|
| last quarter | 6 | 4 | 6 | 5 | 5 | 3 | 3 | 2 | | | | |
| New Moon | 14 | 13 | 14 | 13 | 12 | 11 | 10 | 8 | 7 | 6 | 4 | 4 |
| first quarter | 22 | 20 | 22 | 20 | 19 | 17 | 17 | 15 | 14 | 13 | 12 | 12 |
| Full Moon | 28 | 27 | 28 | 27 | 26 | 25 | 24 | 23 | 22 | 21 | 20 | 20 |
| last quarter | | | | | | | | 31 | 29 | 29 | 27 | 26 |

ECLIPSE OF THE MOON *25 June: partial* — North, Central and South America, eastern Australia
ECLIPSE OF THE SUN *11 June: total* — south Asia, Australia, southwest Pacific; *4 December; annular* — Europe, Africa, northeastern South America

| 1984 | 1 | 2 | 3 | 4 | 5 | 6 | 7 | 8 | 9 | 10 | 11 | 12 |
|---|---|---|---|---|---|---|---|---|---|---|---|---|
| New Moon | 3 | 1 | 2 | 1 | 1 | | | | | | | |
| first quarter | 11 | 10 | 10 | 9 | 8 | 6 | 5 | 4 | 2 | 1 | | |
| Full Moon | 18 | 17 | 17 | 15 | 15 | 13 | 13 | 11 | 10 | 9 | 8 | 8 |
| last quarter | 25 | 23 | 24 | 23 | 22 | 21 | 21 | 19 | 18 | 17 | 16 | 15 |
| New Moon | | | | | 30 | 29 | 28 | 26 | 25 | 24 | 22 | 22 |
| first quarter | | | | | | | | | | 31 | 30 | 30 |

ECLIPSE OF THE SUN *30 May: annular and total* — western Europe, North and Central America, northwest Africa, Greenland;
*22 November: total* — southeast Asia, Australia, south Pacific

| 1985 | 1 | 2 | 3 | 4 | 5 | 6 | 7 | 8 | 9 | 10 | 11 | 12 |
|---|---|---|---|---|---|---|---|---|---|---|---|---|
| Full Moon | 7 | 5 | 7 | 5 | 4 | 3 | 2 | | | | | |
| last quarter | 13 | 12 | 13 | 12 | 11 | 10 | 10 | 8 | 7 | 7 | 5 | 5 |
| New Moon | 21 | 19 | 21 | 20 | 19 | 18 | 17 | 16 | 14 | 14 | 12 | 12 |
| first quarter | 29 | 27 | 29 | 28 | 27 | 25 | 24 | 23 | 21 | 20 | 19 | 19 |
| Full Moon | | | | | | | 31 | 30 | 29 | 28 | 27 | 27 |

ECLIPSE OF THE MOON *4 May: total* — Europe, Africa, Asia, Australia;
*28 October: total* — Europe, Asia, Africa, Australia
ECLIPSE OF THE SUN *19 May: partial* — northeast Asia, Arctic, northern North America; *12 November: total* — southern South America, Antarctic, south Atlantic

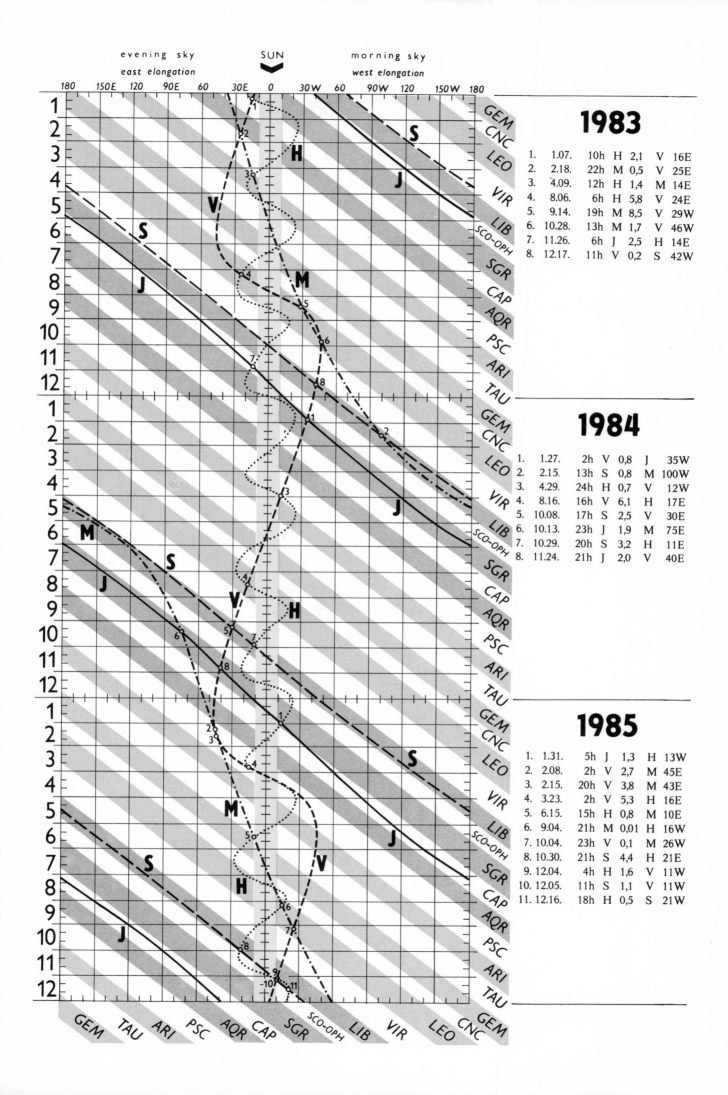

**1983**

| | | | | | | |
|---|---|---|---|---|---|---|
| 1. | 1.07. | 10h | H | 2,1 | V | 16E |
| 2. | 2.18. | 22h | M | 0,5 | V | 25E |
| 3. | 4.09. | 12h | H | 1,4 | M | 14E |
| 4. | 8.06. | 6h | H | 5,8 | V | 24E |
| 5. | 9.14. | 19h | M | 8,5 | V | 29W |
| 6. | 10.28. | 13h | M | 1,7 | V | 46W |
| 7. | 11.26. | 6h | J | 2,5 | H | 14E |
| 8. | 12.17. | 11h | V | 0,2 | S | 42W |

**1984**

| | | | | | | |
|---|---|---|---|---|---|---|
| 1. | 1.27. | 2h | V | 0,8 | J | 35W |
| 2. | 2.15. | 13h | S | 0,8 | M | 100W |
| 3. | 4.29. | 24h | H | 0,7 | V | 12W |
| 4. | 8.16. | 16h | V | 6,1 | H | 17E |
| 5. | 10.08. | 17h | S | 2,5 | V | 30E |
| 6. | 10.13. | 23h | J | 1,9 | M | 75E |
| 7. | 10.29. | 20h | S | 3,2 | H | 11E |
| 8. | 11.24. | 21h | J | 2,0 | V | 40E |

**1985**

| | | | | | | |
|---|---|---|---|---|---|---|
| 1. | 1.31. | 5h | J | 1,3 | H | 13W |
| 2. | 2.08. | 2h | V | 2,7 | M | 45E |
| 3. | 2.15. | 20h | V | 3,8 | M | 43E |
| 4. | 3.23. | 2h | V | 5,3 | H | 16E |
| 5. | 6.15. | 15h | H | 0,8 | M | 10E |
| 6. | 9.04. | 21h | M | 0,01 | H | 16W |
| 7. | 10.04. | 23h | V | 0,1 | M | 26W |
| 8. | 10.30. | 21h | S | 4,4 | H | 21E |
| 9. | 12.04. | 4h | H | 1,6 | V | 11W |
| 10. | 12.05. | 11h | S | 1,1 | V | 11W |
| 11. | 12.16. | 18h | H | 0,5 | S | 21W |

| 1986 | 1 | 2 | 3 | 4 | 5 | 6 | 7 | 8 | 9 | 10 | 11 | 12 |
|---|---|---|---|---|---|---|---|---|---|---|---|---|
| last quarter | 3 | 2 | 3 | 1 | 1 | | | | | | | |
| New Moon | 10 | 9 | 10 | 9 | 8 | 7 | 7 | 5 | 4 | 3 | 2 | 1 |
| first quarter | 17 | 16 | 18 | 17 | 17 | 15 | 14 | 13 | 11 | 10 | 8 | 8 |
| Full Moon | 26 | 24 | 26 | 24 | 23 | 22 | 21 | 19 | 18 | 17 | 16 | 16 |
| last quarter | | | | | 30 | 29 | 28 | 27 | 26 | 25 | 24 | 24 |
| New Moon | | | | | | | | | | | | 31 |

ECLIPSE OF THE MOON *24 April: total* — east Asia, Australia, Antarctic; *17 October: total* — Asia, Europe, Africa, Australia

ECLIPSE OF THE SUN *9 April: partial* — Australia, southern Indian Ocean, part of the Antarctic; *3 October: annular and total* — North and Central America, Greenland

Note: 13 November — transit of Mercury across the solar disc.

| 1987 | 1 | 2 | 3 | 4 | 5 | 6 | 7 | 8 | 9 | 10 | 11 | 12 |
|---|---|---|---|---|---|---|---|---|---|---|---|---|
| first quarter | 6 | 5 | 7 | 6 | 6 | 4 | 4 | 2 | 1 | | | |
| Full Moon | 15 | 13 | 15 | 14 | 13 | 11 | 11 | 9 | 7 | 7 | 5 | 5 |
| last quarter | 22 | 21 | 22 | 20 | 20 | 18 | 17 | 16 | 14 | 14 | 13 | 13 |
| New Moon | 29 | 28 | 29 | 28 | 27 | 26 | 25 | 24 | 23 | 22 | 21 | 20 |
| first quarter | | | | | | | | | 30 | 29 | 28 | 27 |

ECLIPSE OF THE SUN *29 March: annular and total* — eastern South America, part of Africa, south Atlantic; *23 September: annular* — Asia, northeastern Indian Ocean, west Pacific, part of Australia

| 1988 | 1 | 2 | 3 | 4 | 5 | 6 | 7 | 8 | 9 | 10 | 11 | 12 |
|---|---|---|---|---|---|---|---|---|---|---|---|---|
| Full Moon | 4 | 2 | 3 | 2 | 1 | | | | | | | |
| last quarter | 12 | 10 | 11 | 9 | 9 | 7 | 6 | 4 | 3 | 2 | 1 | 1 |
| New Moon | 19 | 17 | 18 | 16 | 15 | 14 | 13 | 12 | 11 | 10 | 9 | 9 |
| first quarter | 25 | 24 | 25 | 23 | 23 | 22 | 22 | 20 | 19 | 18 | 16 | 16 |
| Full Moon | | | | | 31 | 29 | 29 | 27 | 25 | 25 | 23 | 23 |
| last quarter | | | | | | | | | | | | 31 |

ECLIPSE OF THE MOON *3 March: partial* — Asia, eastern Europe, east Africa, Australia; *27 August: partial* — east Asia, Australia, Antarctic, North America

ECLIPSE OF THE SUN *18 March: total* — east Asia, northern Australia, northwestern North America; *11 September: annular* — east Africa, south Asia, Australia, part of the Antarctic

Note: on 22 February 1988 Mars will be only 40″ from Uranus.

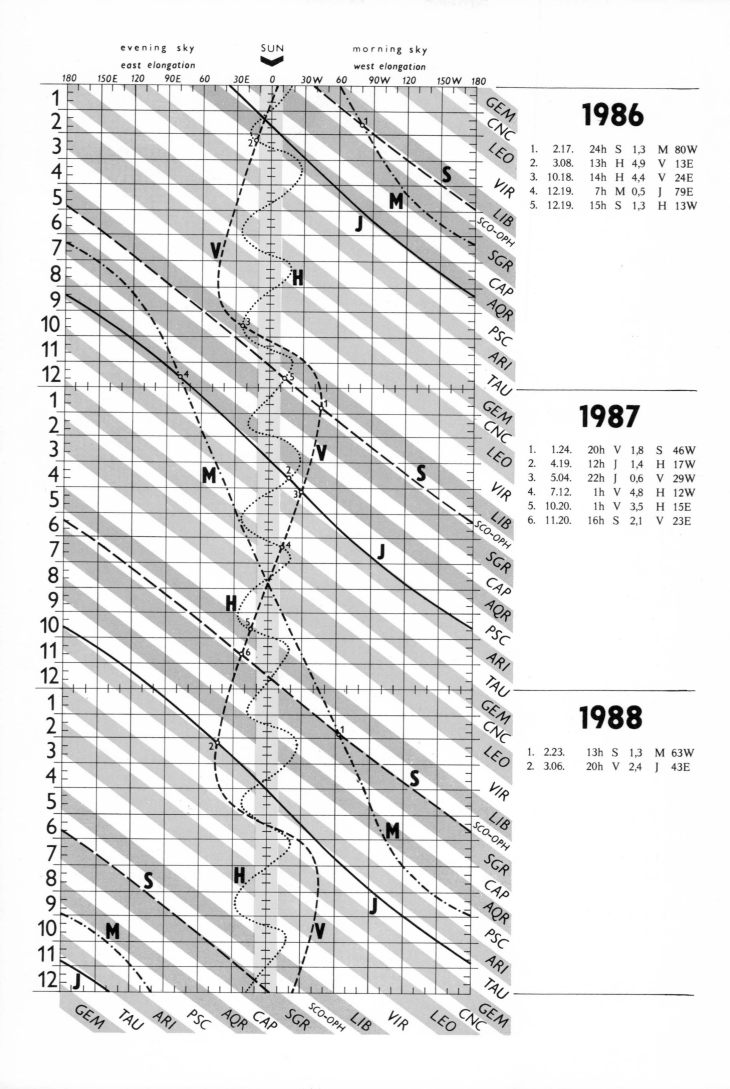

**1986**

1. 2.17. 24h S 1,3 M 80W
2. 3.08. 13h H 4,9 V 13E
3. 10.18. 14h H 4,4 V 24E
4. 12.19. 7h M 0,5 J 79E
5. 12.19. 15h S 1,3 H 13W

**1987**

1. 1.24. 20h V 1,8 S 46W
2. 4.19. 12h J 1,4 H 17W
3. 5.04. 22h J 0,6 V 29W
4. 7.12. 1h V 4,8 H 12W
5. 10.20. 1h V 3,5 H 15E
6. 11.20. 16h S 2,1 V 23E

**1988**

1. 2.23. 13h S 1,3 M 63W
2. 3.06. 20h V 2,4 J 43E

| 1989 | 1 | 2 | 3 | 4 | 5 | 6 | 7 | 8 | 9 | 10 | 11 | 12 |
|---|---|---|---|---|---|---|---|---|---|---|---|---|
| New Moon | 7 | 6 | 7 | 6 | 5 | 3 | 3 | 1 | | | | |
| first quarter | 14 | 12 | 14 | 12 | 12 | 11 | 11 | 9 | 8 | 8 | 6 | 6 |
| Full Moon | 21 | 20 | 22 | 21 | 20 | 19 | 18 | 17 | 15 | 14 | 13 | 12 |
| last quarter | 30 | 28 | 30 | 28 | 28 | 26 | 25 | 23 | 22 | 21 | 20 | 19 |
| New Moon | | | | | | | | 31 | 29 | 29 | 28 | 28 |

ECLIPSE OF THE MOON *20 February: total* — Europe, Asia, Australia, northwestern North America; *17 August: total* — South America, North America, Africa, Europe
ECLIPSE OF THE SUN *7 March: partial* — North America, Greenland, northwest Atlantic; *31 August: partial* — south Africa, southern Indian Ocean, part of the Antarctic

| 1990 | 1 | 2 | 3 | 4 | 5 | 6 | 7 | 8 | 9 | 10 | 11 | 12 |
|---|---|---|---|---|---|---|---|---|---|---|---|---|
| first quarter | 4 | 2 | 4 | 2 | 1 | | | | | | | |
| Full Moon | 11 | 9 | 11 | 10 | 9 | 8 | 8 | 6 | 5 | 4 | 2 | 2 |
| last quarter | 18 | 17 | 19 | 18 | 17 | 16 | 15 | 13 | 11 | 11 | 9 | 9 |
| New Moon | 26 | 25 | 26 | 25 | 24 | 22 | 22 | 20 | 19 | 18 | 17 | 17 |
| first quarter | | | | | 31 | 29 | 29 | 28 | 27 | 26 | 25 | 25 |
| Full Moon | | | | | | | | | | | | 31 |

ECLIPSE OF THE MOON *9 February: total* — Asia, Europe, Africa, Australia, east Atlantic; *6 August: partial* — Australia, Asia, Antarctic
ECLIPSE OF THE SUN *26 January: annular* — Antarctic, southern South America, southwest Atlantic; *22 July: total* — northeastern Europe, north Asia, North America

| 1991 | 1 | 2 | 3 | 4 | 5 | 6 | 7 | 8 | 9 | 10 | 11 | 12 |
|---|---|---|---|---|---|---|---|---|---|---|---|---|
| last quarter | 7 | 6 | 8 | 7 | 7 | 5 | 5 | 3 | 1 | 1 | | |
| New Moon | 15 | 14 | 16 | 14 | 14 | 12 | 11 | 10 | 8 | 7 | 6 | 6 |
| first quarter | 23 | 21 | 23 | 21 | 20 | 19 | 18 | 17 | 15 | 15 | 14 | 14 |
| Full Moon | 30 | 28 | 30 | 28 | 28 | 27 | 26 | 25 | 23 | 23 | 21 | 21 |
| last quarter | | | | | | | | | | 30 | 28 | 28 |

ECLIPSE OF THE MOON *21 December: partial* — northeast Asia, part of Australia, North America, Greenland
ECLIPSE OF THE SUN *15 January: annular* — Australia, part of the Antarctic, south Pacific; *11 July: total* — southern North America, Central America, northern South America, west Atlantic

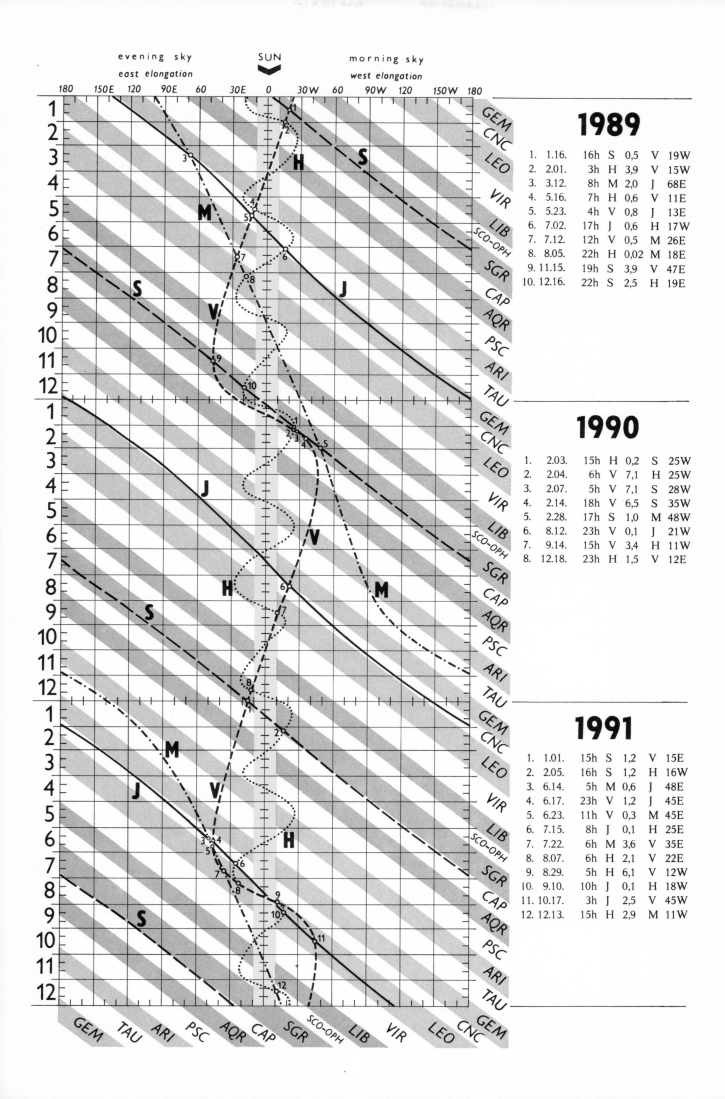

## evening sky
### east elongation

## SUN

## morning sky
### west elongation

| 180 | 150E | 120 | 90E | 60 | 30E | 0 | 30W | 60 | 90W | 120 | 150W | 180 |

# 1989

| | | | | | | | | |
|---|---|---|---|---|---|---|---|---|
| 1. | 1.16. | 16h | S | 0,5 | V | 19W | | |
| 2. | 2.01. | 3h | H | 3,9 | V | 15W | | |
| 3. | 3.12. | 8h | M | 2,0 | J | 68E | | |
| 4. | 5.16. | 7h | H | 0,6 | V | 11E | | |
| 5. | 5.23. | 4h | V | 0,8 | J | 13E | | |
| 6. | 7.02. | 17h | J | 0,6 | H | 17W | | |
| 7. | 7.12. | 12h | V | 0,5 | M | 26E | | |
| 8. | 8.05. | 22h | H | 0,02 | M | 18E | | |
| 9. | 11.15. | 19h | S | 3,9 | V | 47E | | |
| 10. | 12.16. | 22h | S | 2,5 | H | 19E | | |

# 1990

| | | | | | | | |
|---|---|---|---|---|---|---|---|
| 1. | 2.03. | 15h | H | 0,2 | S | 25W |
| 2. | 2.04. | 6h | V | 7,1 | H | 25W |
| 3. | 2.07. | 5h | V | 7,1 | S | 28W |
| 4. | 2.14. | 18h | V | 6,5 | S | 35W |
| 5. | 2.28. | 17h | S | 1,0 | M | 48W |
| 6. | 8.12. | 23h | V | 0,1 | J | 21W |
| 7. | 9.14. | 15h | V | 3,4 | H | 11W |
| 8. | 12.18. | 23h | H | 1,5 | V | 12E |

# 1991

| | | | | | | | |
|---|---|---|---|---|---|---|---|
| 1. | 1.01. | 15h | S | 1,2 | V | 15E |
| 2. | 2.05. | 16h | S | 1,2 | H | 16W |
| 3. | 6.14. | 5h | M | 0,6 | J | 48E |
| 4. | 6.17. | 23h | V | 1,2 | J | 45E |
| 5. | 6.23. | 11h | V | 0,3 | M | 45E |
| 6. | 7.15. | 8h | J | 0,1 | H | 25E |
| 7. | 7.22. | 6h | M | 3,6 | V | 35E |
| 8. | 8.07. | 6h | H | 2,1 | V | 22E |
| 9. | 8.29. | 5h | H | 6,1 | V | 12W |
| 10. | 9.10. | 10h | J | 0,1 | H | 18W |
| 11. | 10.17. | 3h | J | 2,5 | V | 45W |
| 12. | 12.13. | 15h | H | 2,9 | M | 11W |

| 1992 | 1 | 2 | 3 | 4 | 5 | 6 | 7 | 8 | 9 | 10 | 11 | 12 |
|---|---|---|---|---|---|---|---|---|---|---|---|---|
| New Moon | 4 | 3 | 4 | 3 | 2 | 1 | | | | | | |
| first quarter | 13 | 11 | 12 | 10 | 9 | 7 | 7 | 5 | 3 | 3 | 2 | 2 |
| Full Moon | 19 | 18 | 18 | 17 | 16 | 15 | 14 | 13 | 12 | 11 | 10 | 9 |
| last quarter | 26 | 25 | 26 | 24 | 24 | 23 | 22 | 21 | 19 | 19 | 17 | 16 |
| New Moon | | | | | | 30 | 29 | 28 | 26 | 25 | 24 | 24 |

ECLIPSE OF THE MOON *15 June: partial* — South America, North America, Antarctic, west Africa; *9 December: total* — Africa, Europe, Greenland, North America, part of South America, west Asia
ECLIPSE OF THE SUN *4 January: annular* — northeast Australia, western North America; *30 June: total* — South America, south Africa, south Atlantic; *24 December: partial* — northeast Asia, north Pacific

| 1993 | 1 | 2 | 3 | 4 | 5 | 6 | 7 | 8 | 9 | 10 | 11 | 12 |
|---|---|---|---|---|---|---|---|---|---|---|---|---|
| first quarter | 1 | | 1 | | | | | | | | | |
| Full Moon | 8 | 6 | 8 | 6 | 6 | 4 | 3 | 2 | 1 | | | |
| last quarter | 15 | 13 | 15 | 13 | 13 | 12 | 11 | 10 | 9 | 8 | 7 | 6 |
| New Moon | 22 | 21 | 23 | 21 | 21 | 20 | 19 | 17 | 16 | 15 | 13 | 13 |
| first quarter | 30 | | 31 | 29 | 28 | 26 | 26 | 24 | 22 | 22 | 21 | 20 |
| Full Moon | | | | | | | | | 30 | 30 | 29 | 28 |

ECLIPSE OF THE MOON *4 June: total* — east Asia, Australia, Antarctic; *29 November; total* — North and South America, western Europe, northeast Asia
ECLIPSE OF THE SUN *21 May: partial* — North America, Arctic, northern Europe, northwest Asia; *13 November: partial* — south Australia, Antarctic, southern South America
Note: 6 November — transit of Mercury across the solar disc.

| 1994 | 1 | 2 | 3 | 4 | 5 | 6 | 7 | 8 | 9 | 10 | 11 | 12 |
|---|---|---|---|---|---|---|---|---|---|---|---|---|
| last quarter | 5 | 3 | 4 | 3 | 2 | 1 | | | | | | |
| New Moon | 11 | 10 | 12 | 11 | 10 | 9 | 8 | 7 | 5 | 5 | 3 | 2 |
| first quarter | 19 | 18 | 20 | 19 | 18 | 16 | 16 | 14 | 12 | 11 | 10 | 9 |
| Full Moon | 27 | 26 | 27 | 25 | 25 | 23 | 22 | 21 | 19 | 19 | 18 | 18 |
| last quarter | | | | | | 30 | 30 | 29 | 28 | 27 | 26 | 25 |

ECLIPSE OF THE MOON *25 May: partial* — South and North America, Africa, western Europe
ECLIPSE OF THE SUN *10 May: annular* — North and Central America, western Europe, northwest Africa; *3 November: total* — South America, part of the Antarctic, south Africa

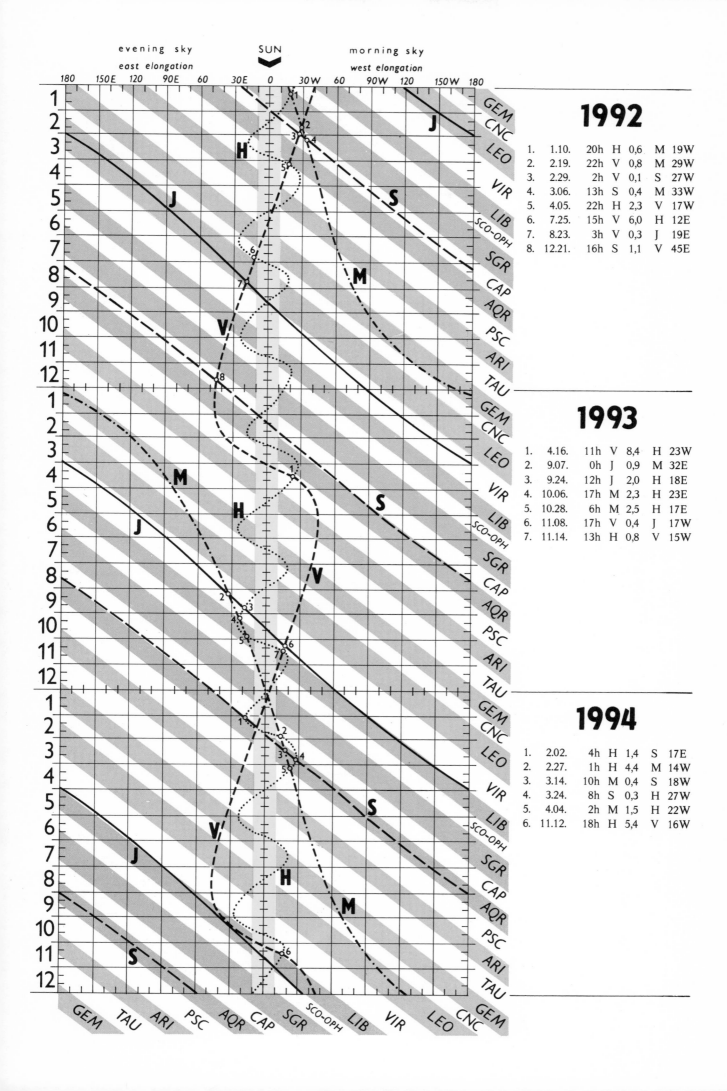

**1992**

| | | | | | | |
|---|---|---|---|---|---|---|
| 1. | 1.10. | 20h | H | 0,6 | M | 19W |
| 2. | 2.19. | 22h | V | 0,8 | M | 29W |
| 3. | 2.29. | 2h | V | 0,1 | S | 27W |
| 4. | 3.06. | 13h | S | 0,4 | M | 33W |
| 5. | 4.05. | 22h | H | 2,3 | V | 17W |
| 6. | 7.25. | 15h | V | 6,0 | H | 12E |
| 7. | 8.23. | 3h | V | 0,3 | J | 19E |
| 8. | 12.21. | 16h | S | 1,1 | V | 45E |

**1993**

| | | | | | | |
|---|---|---|---|---|---|---|
| 1. | 4.16. | 11h | V | 8,4 | H | 23W |
| 2. | 9.07. | 0h | J | 0,9 | M | 32E |
| 3. | 9.24. | 12h | J | 2,0 | H | 18E |
| 4. | 10.06. | 17h | M | 2,3 | H | 23E |
| 5. | 10.28. | 6h | M | 2,5 | H | 17E |
| 6. | 11.08. | 17h | V | 0,4 | J | 17W |
| 7. | 11.14. | 13h | H | 0,8 | V | 15W |

**1994**

| | | | | | | |
|---|---|---|---|---|---|---|
| 1. | 2.02. | 4h | H | 1,4 | S | 17E |
| 2. | 2.27. | 1h | H | 4,4 | M | 14W |
| 3. | 3.14. | 10h | M | 0,4 | S | 18W |
| 4. | 3.24. | 8h | S | 0,3 | H | 27W |
| 5. | 4.04. | 2h | M | 1,5 | H | 22W |
| 6. | 11.12. | 18h | H | 5,4 | V | 16W |

| 1995 | 1 | 2 | 3 | 4 | 5 | 6 | 7 | 8 | 9 | 10 | 11 | 12 |
|---|---|---|---|---|---|---|---|---|---|---|---|---|
| New Moon | 1 | | 1 | | | | | | | | | |
| first quarter | 8 | 7 | 9 | 8 | 7 | 6 | 5 | 4 | 2 | 1 | | |
| Full Moon | 16 | 15 | 17 | 15 | 14 | 13 | 12 | 10 | 9 | 8 | 7 | 7 |
| last quarter | 24 | 22 | 23 | 22 | 21 | 19 | 19 | 18 | 16 | 16 | 15 | 15 |
| New Moon | 30 | | 31 | 29 | 29 | 28 | 27 | 26 | 24 | 24 | 22 | 22 |
| first quarter | | | | | | | | | | 30 | 29 | 28 |

ECLIPSE OF THE MOON *15 April: partial* — east Asia, Australia, western North America, Pacific
ECLIPSE OF THE SUN *29 April: annular* — Central and South America, middle Atlantic, southeast Pacific; *24 October: total* — Asia, north Australia, west Pacific
Note: on 20 May Earth crosses the plane of Saturn's ring from north to south; a second time on 11 August from south to north; on 19 November the Sun crosses the plane of Saturn's ring from north to south.

| 1996 | 1 | 2 | 3 | 4 | 5 | 6 | 7 | 8 | 9 | 10 | 11 | 12 |
|---|---|---|---|---|---|---|---|---|---|---|---|---|
| Full Moon | 5 | 4 | 5 | 4 | 3 | 1 | 1 | | | | | |
| last quarter | 13 | 12 | 12 | 10 | 10 | 8 | 7 | 6 | 4 | 4 | 3 | 3 |
| New Moon | 20 | 18 | 19 | 17 | 17 | 16 | 15 | 14 | 12 | 12 | 11 | 10 |
| first quarter | 27 | 26 | 27 | 25 | 25 | 24 | 23 | 22 | 20 | 19 | 18 | 17 |
| Full Moon | | | | | | | 30 | 28 | 27 | 26 | 25 | 24 |

ECLIPSE OF THE MOON *4 April: total* — Africa, Europe, eastern North America, South America, west Asia; *27 September: total* — North and South America, Africa, Europe
ECLIPSE OF THE SUN *17 April: partial* — South Pacific; *12 October: partial* — northeastern North America, Europe, north Africa
Note: on 12 February Earth crosses the plane of Saturn's ring from north to south.

| 1997 | 1 | 2 | 3 | 4 | 5 | 6 | 7 | 8 | 9 | 10 | 11 | 12 |
|---|---|---|---|---|---|---|---|---|---|---|---|---|
| last quarter | 2 | | 2 | | | | | | | | | |
| New Moon | 9 | 7 | 9 | 7 | 6 | 5 | 4 | 3 | 1 | 1 | | |
| first quarter | 15 | 14 | 16 | 14 | 14 | 13 | 12 | 11 | 10 | 9 | 7 | 7 |
| Full Moon | 23 | 22 | 24 | 22 | 22 | 20 | 20 | 18 | 16 | 16 | 14 | 14 |
| last quarter | 31 | | 31 | 30 | 29 | 27 | 26 | 25 | 23 | 23 | 21 | 21 |
| New Moon | | | | | | | | | | 31 | 30 | 29 |

ECLIPSE OF THE MOON *24 March: partial* — North and South America, west Africa, western Europe; *16 September: total* — Asia, Europe, Africa, Australia, part of the Antarctic
ECLIPSE OF THE SUN *9 March: total* — northeast Asia, northwest Pacific; *2 September: partial* — Australia, southwest Pacific, part of the Antarctic

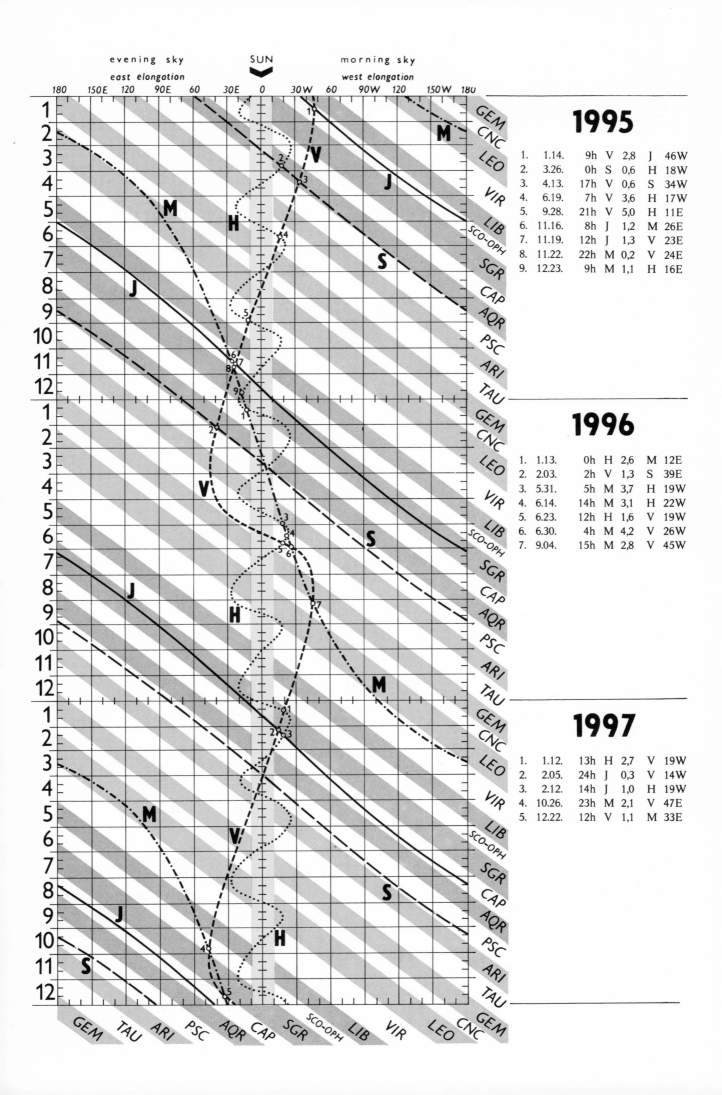

**1995**

| | | | | | | | |
|---|---|---|---|---|---|---|---|
| 1. | 1.14. | 9h | V | 2,8 | J | 46W |
| 2. | 3.26. | 0h | S | 0,6 | H | 18W |
| 3. | 4.13. | 17h | V | 0,6 | S | 34W |
| 4. | 6.19. | 7h | V | 3,6 | H | 17W |
| 5. | 9.28. | 21h | V | 5,0 | H | 11E |
| 6. | 11.16. | 8h | J | 1,2 | M | 26E |
| 7. | 11.19. | 12h | J | 1,3 | V | 23E |
| 8. | 11.22. | 22h | M | 0,2 | V | 24E |
| 9. | 12.23. | 9h | M | 1,1 | H | 16E |

**1996**

| | | | | | | | |
|---|---|---|---|---|---|---|---|
| 1. | 1.13. | 0h | H | 2,6 | M | 12E |
| 2. | 2.03. | 2h | V | 1,3 | S | 39E |
| 3. | 5.31. | 5h | M | 3,7 | H | 19W |
| 4. | 6.14. | 14h | M | 3,1 | H | 22W |
| 5. | 6.23. | 12h | H | 1,6 | V | 19W |
| 6. | 6.30. | 4h | M | 4,2 | V | 26W |
| 7. | 9.04. | 15h | M | 2,8 | V | 45W |

**1997**

| | | | | | | | |
|---|---|---|---|---|---|---|---|
| 1. | 1.12. | 13h | H | 2,7 | V | 19W |
| 2. | 2.05. | 24h | J | 0,3 | V | 14W |
| 3. | 2.12. | 14h | J | 1,0 | H | 19W |
| 4. | 10.26. | 23h | M | 2,1 | V | 47E |
| 5. | 12.22. | 12h | V | 1,1 | M | 33E |

| 1998 | 1 | 2 | 3 | 4 | 5 | 6 | 7 | 8 | 9 | 10 | 11 | 12 |
|---|---|---|---|---|---|---|---|---|---|---|---|---|
| first quarter | 5 | 3 | 5 | 3 | 3 | 2 | 1 | | | | | |
| Full Moon | 12 | 11 | 13 | 11 | 11 | 10 | 9 | 8 | 6 | 5 | 4 | 3 |
| last quarter | 20 | 19 | 21 | 19 | 19 | 17 | 16 | 14 | 13 | 12 | 11 | 10 |
| New Moon | 28 | 26 | 28 | 26 | 25 | 24 | 23 | 22 | 20 | 20 | 19 | 18 |
| first quarter | | | | | | | 31 | 30 | 28 | 28 | 27 | 26 |

ECLIPSE OF THE SUN *26 February: total* — southern North America, Central America, northern South America; *22 August: annular* — southeast Asia, Australia, southwest Pacific

| 1999 | 1 | 2 | 3 | 4 | 5 | 6 | 7 | 8 | 9 | 10 | 11 | 12 |
|---|---|---|---|---|---|---|---|---|---|---|---|---|
| Full Moon | 2 | | 2 | | | | | | | | | |
| last quarter | 9 | 8 | 10 | 9 | 8 | 7 | 6 | 4 | 2 | 2 | | |
| New Moon | 17 | 16 | 17 | 16 | 15 | 13 | 13 | 11 | 9 | 9 | 8 | 7 |
| first quarter | 24 | 23 | 24 | 22 | 22 | 20 | 20 | 19 | 17 | 17 | 16 | 16 |
| Full Moon | 31 | | 31 | 30 | 30 | 28 | 28 | 26 | 25 | 24 | 23 | 22 |
| last quarter | | | | | | | | | | 31 | 29 | 29 |

ECLIPSE OF THE MOON *28 July: partial* — Pacific, east Asia, Australia, Antarctic, northwestern North America
ECLIPSE OF THE SUN *16 February: annular* — south Africa, Antarctic, southeast Asia, Australia; *11 August: total* — eastern North America, Europe, north Africa, west Asia
Note: 15 November — transit of Mercury across the solar disc.

| 2000 | 1 | 2 | 3 | 4 | 5 | 6 | 7 | 8 | 9 | 10 | 11 | 12 |
|---|---|---|---|---|---|---|---|---|---|---|---|---|
| New Moon | 6 | 5 | 6 | 4 | 4 | 2 | 1 | | | | | |
| first quarter | 14 | 12 | 13 | 11 | 10 | 9 | 8 | 7 | 5 | 5 | 4 | 4 |
| Full Moon | 21 | 19 | 20 | 18 | 18 | 16 | 16 | 15 | 13 | 13 | 11 | 11 |
| last quarter | 28 | 27 | 28 | 26 | 26 | 25 | 24 | 22 | 21 | 20 | 18 | 18 |
| New Moon | | | | | | | 31 | 29 | 27 | 27 | 25 | 25 |

ECLIPSE OF THE MOON *21 January: total* — North and South America, west Africa, Europe; *16 July: total* — Australia, east Asia, Antarctic
ECLIPSE OF THE SUN *5 February: partial* — Antarctic, south Atlantic, southern Indian Ocean; *1 July: partial* — southern South America, southeast Pacific; *31 July: partial* — north Asia, Arctic, northwestern North America; *25 December; partial* — North and Central America, northwest Atlantic
Note: 28 March — greatest possible elongation of Mercury: 27° 50′W.

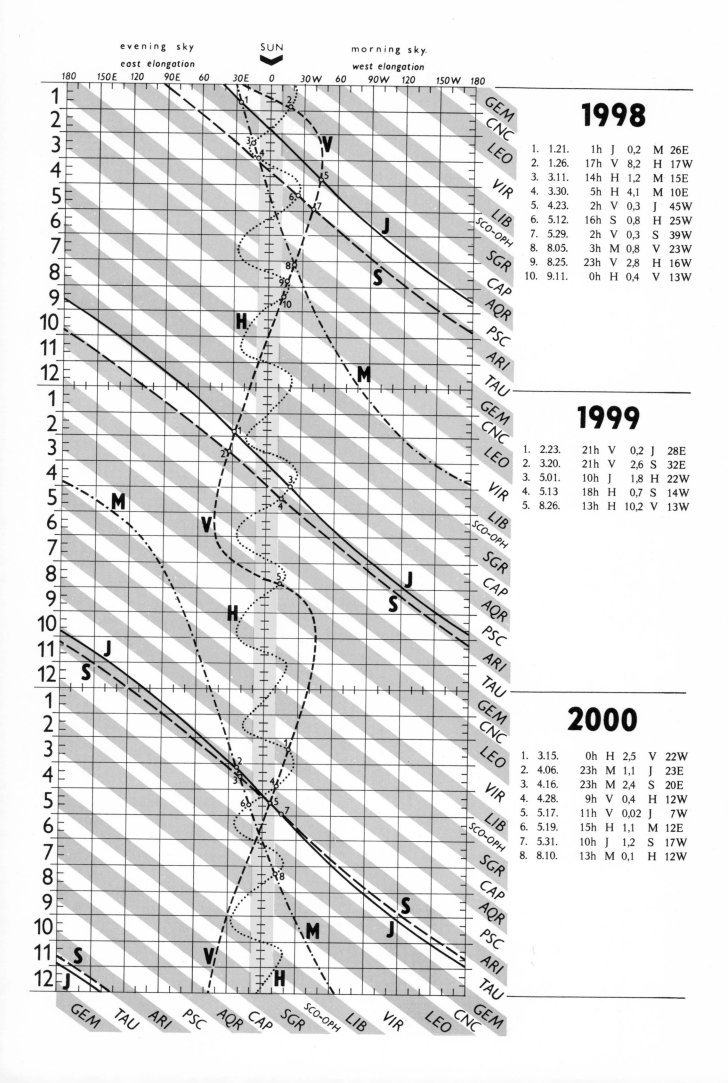

**1998**

| | | | | | | | |
|---|---|---|---|---|---|---|---|
| 1. | 1.21. | 1h | J | 0,2 | M | 26E |
| 2. | 1.26. | 17h | V | 8,2 | H | 17W |
| 3. | 3.11. | 14h | H | 1,2 | M | 15E |
| 4. | 3.30. | 5h | H | 4,1 | M | 10E |
| 5. | 4.23. | 2h | V | 0,3 | J | 45W |
| 6. | 5.12. | 16h | S | 0,8 | H | 25W |
| 7. | 5.29. | 2h | V | 0,3 | S | 39W |
| 8. | 8.05. | 3h | M | 0,8 | V | 23W |
| 9. | 8.25. | 23h | V | 2,8 | H | 16W |
| 10. | 9.11. | 0h | H | 0,4 | V | 13W |

**1999**

| | | | | | | | |
|---|---|---|---|---|---|---|---|
| 1. | 2.23. | 21h | V | 0,2 | J | 28E |
| 2. | 3.20. | 21h | V | 2,6 | S | 32E |
| 3. | 5.01. | 10h | J | 1,8 | H | 22W |
| 4. | 5.13. | 18h | H | 0,7 | S | 14W |
| 5. | 8.26. | 13h | H | 10,2 | V | 13W |

**2000**

| | | | | | | | |
|---|---|---|---|---|---|---|---|
| 1. | 3.15. | 0h | H | 2,5 | V | 22W |
| 2. | 4.06. | 23h | M | 1,1 | J | 23E |
| 3. | 4.16. | 23h | M | 2,4 | S | 20E |
| 4. | 4.28. | 9h | V | 0,4 | H | 12W |
| 5. | 5.17. | 11h | V | 0,02 | J | 7W |
| 6. | 5.19. | 15h | H | 1,1 | M | 12E |
| 7. | 5.31. | 10h | J | 1,2 | S | 17W |
| 8. | 8.10. | 13h | M | 0,1 | H | 12W |

# AMATEUR OBSERVATION

Some amateur astronomers are content to observe the heavens with simple binoculars, whereas others have gone so far as to build an observatory of their own fitted with equipment that even many a university need not be ashamed of. Whatever the instrument, however, it should serve a specific, previously planned observation programme, even if such a programme is merely occasional observation of the sky for one's own pleasure and the enlarging of one's horizons, when even modest equipment will serve the purpose well.

The first thing an amateur usually wants is his own telescope. Whether to buy it in a shop or make it yourself depends on the state of your finances and how clever you are with your hands. What type of telescope should we choose?

In substance we have a choice of two basic types: a *refractor* in which the principle focusing element is an achromatic lens, or a *reflector*, in which the principal focusing element is a mirror. The principles of the two systems are shown in Figure 4. In a refractor (Figure 4, left) light rays pass through the lens $O_1$ and converge to a focus, forming a small image in the focal plane; this is then magnified by and viewed through the eyepiece $O_2$, the degree of magnification being determined by the choice of eyepiece. In a reflector (Figure 4, right) light rays are reflected by a concave mirror $O_1$ (objective) usually onto a secondary mirror which directs the

beam to the eyepiece $O_2$. The objective consists of a very precisely ground spherical or paraboloid surface which is coated with silver or more commonly with aluminium so as to reflect light with the greatest possible efficiency. There are several types of reflectors, differing in the arrangement and number of optical components and in the shape of the mirrors; Figure 4 shows a schematic diagram of the so-called Newton reflector. A type that is frequently used is the Cassegrain reflector with hole in the axis of the primary mirror; the secondary mirror is then convex and reflects the beam into the hole in the primary mirror behind which is the eyepiece. A good practical system is the Cassegrain telescope fitted with a correcting plate (meniscus lens) that eliminates the optical aberrations of a simple mirror objective. This is the principle used in the portable telescopes available under brand names such as Questar, Celestron, Meniscus-Cassegrain.

What is better — a refractor or reflector? There is no clear-cut answer to this question. Factors determining the choice include the suitability of the given type for the given purpose (for observation of the Sun, planets or comets, perhaps), the cost, and above all the quality of the optical system of the given instrument. Reflectors are popular with amateurs because they are less expensive and easily accessible. With skill and patience an amateur can even grind a quality mirror with a diameter of 10 to 20 centimetres or more by himself under the guidance of an expert. On the other hand, the objective lens of a refractor is usually double-convex, of special types of optical glass, and cannot be made in the home workshop.

Constructing a simple refractor from an eyeglass, magnifying glass and paper tube can be done easily and quickly even by a small schoolboy. Building a perfectly functioning telescope with objective of 10 or more centimetres in diameter, however, places great demands on both the construction

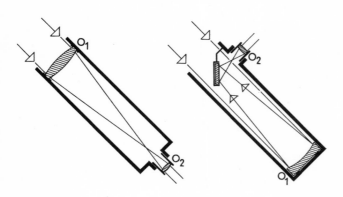

*Fig. 4. Refractor and reflector*

of the lens system and the assembly which holds the lenses in place.

Just as important as the optical system is the mounting of the telescope. This not only serves for pointing the telescope in whatever direction is desired, but must also assure adequate stability of the instrument and exclude the possibility of undesirable vibrations. Light, portable tripods on which the telescope vibrates at the slightest touch of the observer or breath of wind are definitely ruled out. However, even a heavy, massive tripod is no better if the telescope is not properly attached. Even when observing the sky with binoculars, it is desirable to eliminate vibrations; this is easily demonstrated by looking at the Moon through binoculars held in the hand and through the same binoculars attached to a tripod or at least placed against the windowframe for support.

Mountings of astronomical telescopes are of two kinds: azimuthal and parallactic. In both instances the telescope can be rotated round two axes perpendicular to each other. *Azimuthal mounting* (Figure 5, right) has a vertical axis and a horizontal axis. It can be easily constructed and is best suited for small instruments. More suitable, of course, is *parallactic mounting,* where one axis points towards the celestial pole P and is parallel to the Earth's axis of rotation; this axis (called the polar axis) therefore lies in the plane of the local meridian and with the horizontal plane forms an angle equal to the geographical latitude of the place of observation φ (phi). With the aid of a clockwork drive or electric motor the telescope can then automatically track the observed object by rotating round one axis. In addition, parallactic mountings as a rule also have graduated circles for finding objects according to equatorial coordinates. A good mounting should also include mechanical equipment for fine motion of the telescope, for exact adjustment of the object in the field of view, and the like.

What performance can we expect of an amateur telescope? Assuming good quality optics, stable mounting and excellent conditions of observation (a calm and clear night), such a telescope can resolve two points (two stars, two details on the surface of a planet) with an angular separation of $r'' = \frac{120''}{D}$,

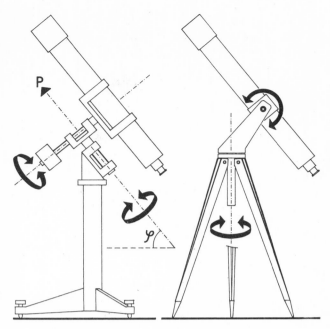

*Fig. 5. Parallactic and azimuthal mounting*

where D is the diameter of the objective in millimetres. For example, an objective 60 millimetres in diameter should theoretically resolve double stars with a distance of 2″ or more between the two components. The actual resolving power of a telescope is best tested by observation of double stars with known separation of components. Much depends also on the keenness of the observer's eyesight and on his experience.

The diameter of the objective is a determining factor not only as regards resolving power but also for the limit of stellar magnitude m visible with the telescope. For example where D = 60 mm, m is 11, where D = 100 mm, m is 12, and so on. The actual value m may be tested, for example, by observing a selected area of stars with known stellar magnitudes. The so-called north polar sequence — a selected area of stars near the pole and thus always visible — is ideal for this purpose (see Figure 6 and the adjacent list of stellar magnitudes on p. 178).

The magnification of the telescope n is determined by the ratio of the focal length of the objective F to the focal length of the eyepiece f:

Magnification can be changed according to need by a suitable choice of various eyepieces. If, for example, we have an objective

177

*Fig. 6. North polar sequence*

Table of fotovisual stellar magnitudes of selected stars.

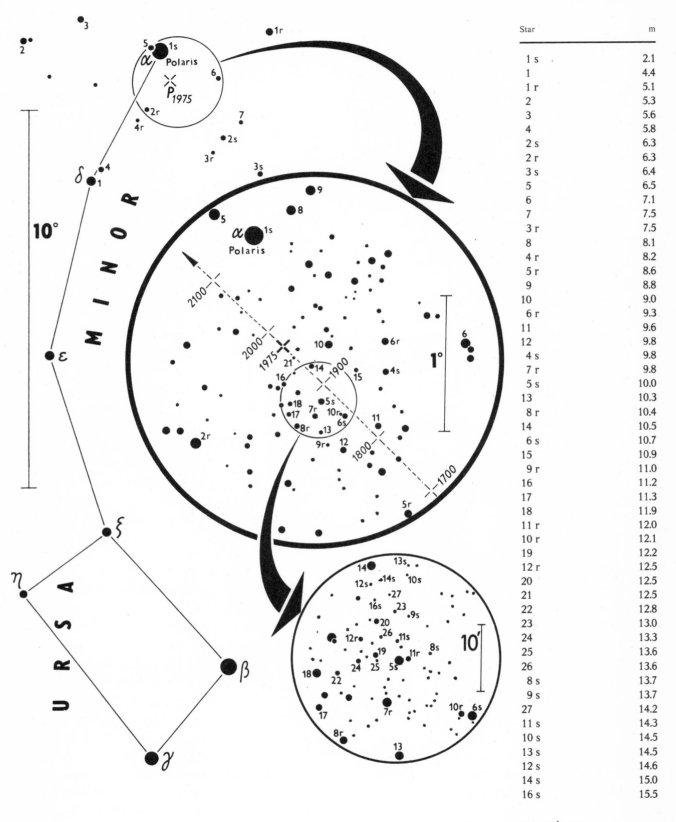

| Star | m |
|---|---|
| 1 s | 2.1 |
| 1 | 4.4 |
| 1 r | 5.1 |
| 2 | 5.3 |
| 3 | 5.6 |
| 4 | 5.8 |
| 2 s | 6.3 |
| 2 r | 6.3 |
| 3 s | 6.4 |
| 5 | 6.5 |
| 6 | 7.1 |
| 7 | 7.5 |
| 3 r | 7.5 |
| 8 | 8.1 |
| 4 r | 8.2 |
| 5 r | 8.6 |
| 9 | 8.8 |
| 10 | 9.0 |
| 6 r | 9.3 |
| 11 | 9.6 |
| 12 | 9.8 |
| 4 s | 9.8 |
| 7 r | 9.8 |
| 5 s | 10.0 |
| 13 | 10.3 |
| 8 r | 10.4 |
| 14 | 10.5 |
| 6 s | 10.7 |
| 15 | 10.9 |
| 9 r | 11.0 |
| 16 | 11.2 |
| 17 | 11.3 |
| 18 | 11.9 |
| 11 r | 12.0 |
| 10 r | 12.1 |
| 19 | 12.2 |
| 12 r | 12.5 |
| 20 | 12.5 |
| 21 | 12.5 |
| 22 | 12.8 |
| 23 | 13.0 |
| 24 | 13.3 |
| 25 | 13.6 |
| 26 | 13.6 |
| 8 s | 13.7 |
| 9 s | 13.7 |
| 27 | 14.2 |
| 11 s | 14.3 |
| 10 s | 14.5 |
| 13 s | 14.5 |
| 12 s | 14.6 |
| 14 s | 15.0 |
| 16 s | 15.5 |

r — red stars
s — supplementary stars
The dates in Figure 6 mark the positions of the North celestial pole.

with focal length F = 1000 mm, then an eyepiece with focal length f = 40 mm will give 25-fold magnification, one with f = 20 mm 50-fold magnification, and one with f = 10 mm 100-fold magnification. Magnification in itself does not tell us much about the quality of a telescope. There is no sense in using extremely large magnification. Each telescope has its own level of maximum useful magnification, and although the *image* can be magnified over and above that, it will yield no new or further *details*. The maximum useful magnification is usually about the equivalent of the diameter of the objective expressed in millimetres (about 100-fold magnification for a 100 mm objective). The magnification which can actually be used is naturally determined also by the conditions of observation, particularly by atmospheric disturbances.

The telescope's *field of view* is determined by the magnification and type of eyepiece. The larger the magnification the smaller the section of sky visible in the field of view.

An astronomical telescope can be used not only for visual observation, but also for photography. Photographs taken in the focal plane of the telescope's objective are common. By attaching a camera without a lens to the telescope in place of the eyepiece, we get the equivalent of a camera with large telephoto lens that can be used, for example, to photograph a solar eclipse, lunar eclipse, or surface of the Moon in various phases. When photographing faint objects such as star clusters and nebulae, it is necessary to count with long exposures (minutes, tens of minutes and even whole hours). For this purpose a good parallactic mounting is a must, as is careful tracking of the object, best of all with a clockwork drive and an occasional visual check with an auxiliary telescope fitted with an illuminated cross-wire in the field of view of the eyepiece.

Good photographs of the starry sky, entire constellations, the Milky Way, interesting configurations of the planets and Moon, meteors, comets, polar auroras, and so on, can easily be made even with an ordinary 35 mm or roll film camera. With an exposure of more than several seconds the daily rotation of the sky will appear on the film by

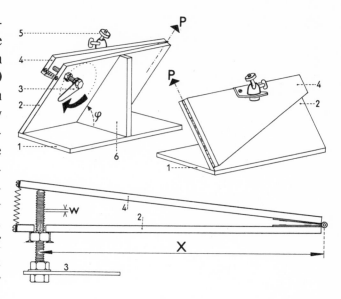

*Fig. 7. Simple parallactic mounting for a camera*

elongation of the images of the celestial objects. If we wish the stars to be depicted as points even on long-exposure photographs, then we mus use a parallactic mounting so that the camera will turn with the sky.

How a simple and easily moved parallactic mounting for a camera can be made by the amateur is shown in Figure 7. Attach plate 2 to base plate 1 using a triangular piece of wood so that the two form an angle equal to the local geographical latitude φ. Attached to plate 2 by means of a hinge is plate 4 carrying on its upper edge a ball tripod-head for securing the camera. The axis of the 'hinge' must point towards the celestial pole when photographing — check this with a compass or according to the Pole Star. Firmly attached to plate 2 is a nut in which is fitted a bolt with rounded end touching plate 4.

Distance X between the axis of the bolt and axis of the hinge should be such that one turn of the bolt, the height of thread W, corresponds to an angular rotation of plate 4 (together with the camera) of 1 minute of time or 0.25 degree of arc.

$X = W : (1.00274 \times \tan 0.25°) = W \times 228.5$
For example, if W = 1 mm, then X = = 228.5 mm. In practice all that needs to be done then during exposure is to turn arm 3 concurrently with the movement of the sec-

ond hand of a watch. A handy do-it-yourselfer will have no difficulty 'motorizing' this single movement. Then a several-minute exposure made possible by such a mounting is quite sufficient even for capturing on film stars that cannot be seen with the unaided eye.

Finally, a few more words to the novice who would like to make expert, systematic and scientifically valuable observations. With the development of science the selection of experimental research programmes for the amateur scientist is becoming increasingly difficult; but in astronomy there are still fields where the cooperation of the amateur is greatly welcome. Such contributions, for example, may be the occasional observation of bolides, comets, new stars, eclipses of the Sun or Moon, or systematic observation of variable stars, meteors or the occultations of stars by the Moon. In some cases even very simple equipment will serve the purpose.

No matter what, however, observations should always be performed purposefully, according to specific rules and a precisely determined programme, with all the essentials required for the final evaluation. Careful records should be made of every observation; what such a record contains depends on the kind of observation. In most cases it will be such items as the place of observation (sometimes it is necessary to know the exact geographical coordinates), the instruments used (parameters of the telescope, magnification used, and so on), data about the photographic apparatus (optical system, type of film, exposure), sometimes also meteorological data, and the like. First and foremost, however, it is always necessary to record the *time* of observation with the required accuracy.

Keeping track of accurate time should be a matter of course for every amateur astronomer. Except for more exact measurements requiring a precision of tenths to hundredths of a second, a good wristwatch, set regularly according to the radio time-signal, will serve the purpose well. Every watch (even an expensive quartz wristwatch) goes a bit fast or slow. It is naturally desirable that such daily deviations be as small as possible, say only a few seconds, but far more important as regards the quality of a watch is its

regularity. Time signals broadcast every hour over the radio for the general public (for scientific purposes many stations broadcast special signals) enable us to find a correction of watches exactly shortly before observation and after and thus obtain a precision of one second without any great difficulty.

Amateur astronomers throughout the world form associations that arrange the necessary contacts with the astronomical community. We therefore strongly recommend that all who are interested in contributing their bit do so in cooperation with other friends of astronomy.

In the UK the best way to start is by joining the Junior Astronomical Society, an organization for 'junior astronomers' — that is complete beginners — not just for young people. Their address is 58 Vaughan Gardens, Ilford, Essex IG1 2PD. With more experience, many amateurs move on to join the British Astronomical Association, which can be contacted at Burlington House, London Wl. Through the JAS and the BAA, amateurs in Britain can make considerable contributions to scientific programmes of astronomical observation.

In the US the Association of Lunar and Planetary Observers (Box AZ, University Park, New Mexico) and the Astronomical Society of the Pacific (675 18th Av., San Francisco, California) deserve special mention, while the names and addresses of many local societies can, of course, be obtained from the appropriate 'phone book.

In Canada, the best way to make initial contacts is through the Royal Astronomical Society of Canada, and similarly the appropriate Royal Astronomical Societies in other Commonwealth countries will always be happy to inform the serious amateur of the existence of groups of observers working in particular areas (such as comet studies), as well as offering the appropriate services to their professional members. A letter (*not* a personal visit or 'phone call) to one of the great observatories will usually elicit a helpful response from professional astronomers happy to help a beginner by putting him or her in contact with other amateurs — after all, that is exactly the way many of the professionals started in astronomy themselves.

*Index to maps of the Moon*
*(maps L1—L6 on pp. 47—57)*

## MOUNTAIN RANGES AND FAULTS

**a** Montes Alpes (Alps) *2*
**b** Montes Apenninus (Apennines) *2*
**c** Montes Carpatus (Carpathians) *1, 2*
**d** Montes Caucasus (Caucasus) *2*
**e** Montes Cordillera (Cordilleras) *4*
**f** Montes Jura *1, 2*
**g** Montes Riphaeus *4*
**h** Montes Spitzbergensis (Spitzbergen) *2*
**k** Montes Taurus *3*
**m** Rupes Altai *5, 6*
**n** Rupes Cauchy *3*
**p** Rupes Recta (Straight Wall) *5*

## MOUNTAINS AND VALLEYS

**r** Piton *2*
**s** Rümker *1*
**t** Vallis Alpes *2*
**u** Vallis Rheita *6*
**v** Vallis Schröteri *1*

## SEAS

**A** Mare Australe (Southern Sea) *6*
**Co** Mare Cognitum (Known Sea) *4, 5*
**Cr** Mare Crisium (Sea of Crises) *3*
**Fe** Mare Fecunditatis (Sea of Fertility) *3, 6*
**Fr** Mare Frigoris (Sea of Cold) *2, 3*
**Hm** Mare Humboldtianum (Humboldt's Sea) *3*
**Hu** Mare Humorum (Sea of Moisture) *4*
**Im** Mare Imbrium (Sea of Rains) *1, 2*
**Ma** Mare Marginis (Border Sea) *3*
**Ne** Mare Nectaris (Sea of Nectar) *6*
**Nu** Mare Nubium (Sea of Clouds) *4, 5*
**O.P.** Oceanus Procellarum (Ocean of Storms) *1, 4*
**Se** Mare Serenitatis (Sea of Serenity) *2, 3*
**Sm** Mare Smythii (Smyth's Sea) *3, 6*
**Sp** Mare Spumans (Foaming Sea) *3, 6*
**Tr** Mare Tranquillitatis (Sea of Tranquillity) *3*
**Un** Mare Undarum (Sea of Waves) *3*
**Va** Mare Vaporum (Sea of Vapours) *2*

## LAKES, MARSHES AND BAYS

**L.M.** Lacus Mortis (Lake of Death) *2, 3*
**L.S.** Lacus Somniorum (Lake of Dreams) *2, 3*
**S.A.** Sinus Aestuum (Bay of Billows) *2*
**S.I.** Sinus Iridum (Bay of Rainbows) *1, 2*
**S.M.** Sinus Medii (Central Bay) *2, 5*
**S.R.** Sinus Roris (Bay of Dew) *1*
**P.E.** Palus Epidemiarum (Marsh of Epidemics) *4*
**P.P.** Palus Putredinis (Marsh of Putrefaction) *2*
**P.S.** Palus Somni (Marsh of Sleep) *3*

## LUNAR PROBES

*whose landing sites are marked on maps L 1 — L 6.*

L   Luna (USSR)  
S   Surveyor (USA)  
R   Ranger (USA)  
A   Apollo (USA)

| Probe | Date | Map | Result |
|---|---|---|---|
| **Luna 2** | 13.  9. 1959 | 2 | First probe to reach the Moon |
| **Ranger 7** | 31.  7. 1964 | 3, 4 | 4,308 close photographs of the Moon, details down to 1 m |
| **Ranger 8** | 20.  2. 1965 | 3, 6 | 7,137 photographs of the Moon |
| **Ranger 9** | 24.  3. 1965 | 5 | 5,814 photographs, showing 25 cm details |
| **Luna 9** | 3.  2. 1966 | 1 | First soft landing; 4 panoramic photographs |
| **Surveyor 1** | 2.  6. 1966 | 1, 4 | soft landing; 11,240 photographs |
| **Luna 13** | 24. 12. 1966 | 1 | 3 panoramic photographs, mechanical soil-sampler |
| **Surveyor 3** | 20.  4. 1967 | 4 | 6,326 photographs; later Apollo 12 landed here |
| **Surveyor 5** | 11.  9. 1967 | 3, 6 | 19,118 photographs; tests of properties of lunar surface |
| **Surveyor 6** | 10. 11. 1967 | 2, 5 | 29,952 photographs |
| **Surveyor 7** | 10.  1. 1968 | 5 | 21,038 photographs; mechanical scoop, chemical analyses |
| **Apollo 11** | 20.  7. 1969 | 3, 6 | first men on the Moon: Armstrong, Aldrin; Collins in orbit |
| **Apollo 12** | 19. 11. 1969 | 4 | Conrad and Bean on the Moon, Gordon in orbit |
| **Luna 16** | 21.  9. 1970 | 3, 6 | automatic collection of a lunar rock sample, brought back to Earth |
| **Luna 17** | 17. 11. 1970 | 1 | mobile laboratory Lunokhod 1 (travelled 10,540 metres) |
| **Apollo 14** | 5.  2. 1971 | 4, 5 | Shepard and Mitchell on the Moon, Roosa in orbit |
| **Apollo 15** | 30.  7. 1971 | 2 | Scott and Irwin on the Moon, Worden in orbit |
| **Luna 20** | 21.  2. 1971 | 3 | automatic collection of a lunar rock sample, brought back to Earth |
| **Apollo 16** | 21.  4. 1972 | 5 | Young and Duke on the Moon, Mattingly in orbit |
| **Apollo 17** | 11. 12. 1972 | 3 | Cernan and Schmitt on the Moon, Evans in orbit |
| **Luna 21** | 15.  1. 1973 | 3 | mobile laboratory Lunokhod 2 |
| **Luna 24** | 18.  8. 1976 | 3 | automatic collection of a lunar rock sample to a depth of 2 metres |

## THE GREEK ALPHABET

| | |
|---|---|
| α | alpha |
| β | béta |
| γ | gamma |
| δ | delta |
| ε | epsilon |
| ζ | zéta |
| η | eta |
| ϑ | theta |
| ι | iota |
| κ | kappa |
| λ | lambda |
| μ | mu |
| ν | nu |
| ξ | xi |
| ο | omicron |
| π | pi |
| ρ | rhó |
| σ | sigma |
| τ | tau |
| υ | ypsilon |
| φ | phi |
| χ | chi |
| ψ | psi |
| ω | omega |

## THE CONSTELLATIONS

| | | | | | |
|---|---|---|---|---|---|
| Andromeda | And | Andromeda | Centaurus | Cen | The Centaur |
| Antlia | Ant | The Air Pump | Cepheus | Cep | Cepheus |
| Apus | Aps | The Bird of Paradice | Cetus | Cet | The Whale |
| Aquarius | Aqr | The Water Carrier | Chamalleon | Cha | The Chameleon |
| Aquila | Aql | The Eagle | Circinus | Cir | The Pair of Compasses |
| Ara | Ara | The Altar | Columba | Col | The Dove |
| Aries | Ari | The Ram | Coma Berenices | Com | Berenice's Hair |
| Auriga | Aur | The Charioteer | Corona Australis | CrA | The Southern Crown |
| Boötes | Boo | The Herdsman | Corona Borealis | CrB | The Northern Crown |
| Caelum | Cae | The Graving Tool | Corvus | Crv | The Crow |
| Camelopardalis | Cam | The Giraffe | Crater | Crt | The Cup |
| Cancer | Cnc | The Crab | Crux | Cru | The Cross |
| Canes Venatici | CVn | The Hunting Dogs | Cygnus | Cyg | The Swan |
| Canis Major | CMa | The Greater Dog | Delphinus | Del | The Dolphin |
| Canis Minor | CMi | The Lesser Dog | Dorado | Dor | The Goldfish |
| Capricornus | Cap | The Goat | Draco | Dra | The Dragon |
| Carina | Car | The Keel | Equuleus | Equ | The Foal |
| Cassiopeia | Cas | Cassiopeia | Eridanus | Eri | The River Eridanus |
| | | | Fornax | For | The Furnace |
| | | | Gemini | Gem | The Twins |
| | | | Grus | Gru | The Crane |

| Name | Abbr. | English |
|---|---|---|
| Hercules | Her | Hercules |
| Horologium | Hor | The Pendulum Clock |
| Hydra | Hya | The Water Snake |
| Hydrus | Hyi | The Lesser Water Snake |
| Indus | Ind | The Indian |
| Lacerta | Lac | The Lizard |
| Leo | Leo | The Lion |
| Leo Minor | LMi | The Lesser Lion |
| Lepus | Lep | The Hare |
| Libra | Lib | The Scales |
| Lupus | Lup | The Wolf |
| Lynx | Lyn | The Lynx |
| Lyra | Lyr | The Lyre |
| Mensa | Men | The Table Mountain |
| Microscopium | Mic | The Microscope |
| Monoceros | Mon | The Unicorn |
| Musca | Mus | The Fly |
| Norma | Nor | The Level |
| Octans | Oct | The Octant |
| Ophiuchus | Oph | The Serpent Holder |
| Orion | Ori | Orion |
| Pavo | Pav | The Peacock |
| Pegasus | Peg | Pegasus |
| Perseus | Per | Perseus |
| Phoenix | Phe | The Phoenix |
| Pictor | Pic | The Painter's Easel |
| Pisces | Psc | The Fishes |
| Piscis Austrinus | PsA | The Southern Fish |
| Puppis | Pup | The Stern |
| Pyxis | Pyx | The Mariner's Compass |
| Reticulum | Ret | The Net |
| Sagitta | Sge | The Arrow |
| Sagittarius | Sgr | The Archer |
| Scorpius | Sco | The Scorpion |
| Sculptor | Scl | The Sculptor |
| Scutum | Sct | The Shield |
| Serpens — Caput and Serpent — Cauda | Ser | The Serpent |
| Sextans | Sex | The Sextant |
| Taurus | Tau | The Bull |
| Telescopium | Tel | The Telescope |
| Triangulum | Tri | The Triangle |
| Triangulum Australe | TrA | The Southern Triangle |
| Tucana | Tuc | The Toucan |
| Ursa Major | UMa | The Great Bear |
| Ursa Minor | UMi | The Little Bear |
| Vela | Vel | The Sail |
| Virgo | Vir | The Virgin |
| Volans | Vol | The Flying Fish |
| Vulpecula | Vul | The Fox |

# GENERAL INDEX